T0260920

Fermented Landscapes

Fermented Landscapes

Lively Processes of
Socio-environmental
Transformation

Edited and with an introduction by
Colleen C. Myles

University of Nebraska Press   LINCOLN

© 2020 by the Board of Regents of the
University of Nebraska

All rights reserved

Library of Congress Cataloging-in-
Publication Data
Names: Myles, Colleen C., editor, writer of
introduction.
Title: Fermented landscapes: lively processes
of socio-environmental transformation /
edited and with an introduction by Colleen
C. Myles.
Description: Lincoln: University of
Nebraska Press, [2020] | Includes
bibliographical references and index.
Identifiers: LCCN 2019042811
ISBN 9781496207760 (hardback)
ISBN 9781496219893 (epub)
ISBN 9781496219909 (mobi)
ISBN 9781496219916 (pdf)
Subjects: LCSH: Alcoholic beverage
industry—Environmental aspects. |
Ecological disturbances. |
Landscape changes.
Classification: LCC HD9350.5 .F47
2020 | DDC 338.4/76631—dc23
LC record available at
https://lccn.loc.gov/2019042811

Set in Garamond Premier Pro
by Mikala R. Kolander.
Designed by L. Auten.

For my children—Orrin, Eleanor, and Iris

Contents

Illustrations

Tables

Acknowledgments

This book is the result of several years of thought, conversation, reflection, and effort. I began this project in early 2016, when I was invited to give a colloquium lecture. I thought, "What better time than the present to try out some of the ideas that have been swirling around in my head related to landscape change and fermentation?" From that beginning I began discussing my evolving thoughts on this subject at essentially every colloquium and conference in which I participated. By early 2017 I knew this project would be best served if pursued as a collective effort, so I secured a book contract for an edited volume and began inviting contributors. I asked my collaborators, colleagues, and trusted intellectual partners to think through what a conceptualization of "fermented landscapes" might look like. I am proud to say that those whose words appear here—and many others who did not provide a chapter—were instrumental in the development and completion of this work. Together this cohort of scholars has outlined, detailed, and explored what it means to think of the material and metaphorical dimensions of fermentation at a variety of scales, from the landscape level all the way into the body.

This book as a whole takes a comprehensive look at complex processes of fermentation to consider the macro consequences of micro(be) processes of socio-environmental transformation. And it simply would not have been possible without the hard work and dedication of the contributors and a number of other key individuals.

I would first like to formally thank those who talked through these ideas with me, both in professional settings and over fermented beverages in other contexts. Many of those with whom I discussed these ideas ended up with chapters in the book, but some did not. These latter folks include (in no particular order) Jessi Breen, Delorean Wiley, Colin Iliff, Garrett Wolf, Jacob Wolfe, Madison Pevey, Trina Filan, Yvon Le Caro, Ashley Jenkins, Sasha Savelyev, Paepin Goff, and surely others. Some people whose essays do appear in the volume but whose influence is underrepresented by the amount of space they occupy there include Ryan Galt, Innisfree McKinnon, Eric Sarmiento, Walter W. Furness, Christopher R. Holtkamp, and Vaughn Bryan Baltzly.

I would also like to thank Bridget Barry of the University of Nebraska Press for her enthusiasm and ongoing support as this project lumbered along. Her advice and feedback—as well as that of two anonymous, external peer reviewers—were indispensable as I worked to weave the pieces of this book, which is interdisciplinary and purposefully wide-ranging in its considerations, into a coherent whole.

The assistance of my graduate research assistant, Shadi Maleki, was similarly indispensable. From sitting with me in my office as I thought through how to proceed, to organizing files, to formatting and format-ting and formatting some more, to summarizing and compiling, to map-making, to working late into the night to meet deadlines, Shadi truly was a lifesaver and I cannot thank her enough.

Having long recognized the importance of a healthy work/life bal-ance, I can attest to how essential a caring, compassionate personal net-work is to (my own) professional success. I thus want to acknowledge the contributions of those who make up my personal support system. Thank you to my dearest friends—Jennifer Devine, Geneva Gano, Kristy Daniel, Mar Huertas, Sarah Fritts, Camila Carlos, and Blake and Brittaney Viebrock—who make all the toil both fun and worth-while, especially when we talk about the details of fermentation over a well-crafted fermented beverage. And of course I could not possibly be who I am meant to be without the guidance and empathy of my fam-ily. My sincerest gratitude goes out to my parents—Charlie Robinson and especially my mother, Laura Robinson, who is always 100 percent supportive and enthusiastic, no matter what I am working on; my zister, Kris Rippee, who has been an advocate for my writing since childhood;

my children, who have been patient and kind and, most of all, loving; and to my spouse, Vaughn Bryan Myles Baltzly, who continues to be a steadfast partner to me in all things.

I could not do this work without the care, attention, and dedication of all these people and many more. Any and all remaining errors or omissions are solely my own.

Introduction

Colleen C. Myles

Fermentation is a state of agitation, turbulent change, or development and is most often associated with the process wherein an agent causes an organic substance to break down into simpler substances. While microorganisms like yeast or bacteria are the usual actors in the biochemical process of fermentation, the (micro)organisms and (f)actors involved in the processes of social, cultural, and political fermentation are less clear. This book explores shifting patterns of land use and management as well as cultural changes related to the production and consumption of fermented beverages in a variety of contexts—a research area I call "fermented landscapes." Specifically, through a variety of material-semiotic analyses, the chapters within this volume take a comprehensive look at complex processes of fermentation to consider the macro consequences of micro(be) processes of socio-environmental transformation. The chapters investigate how the framework of fermented landscapes can be used to examine the environmental, economic, and sociocultural implications of fermentation in both expected and unexpected places and ways.

This book is divided into three parts. The first one deals with conceptualizing the role of fermentation in processes of landscape change; the second addresses landscapes of ferment, alcoholic or otherwise; and the third offers perspectives on the possibilities and limitations of linking fermentation and landscape.

Part 1 frames the concept of fermented landscapes. Chapter 1 outlines the impetus and analytical framework of the concept. Chapter 2, which I wrote with coauthors Christopher Holtkamp, Innisfree McKinnon, Vaughn Bryan Baltzly, and Colton Coiner, considers "booze as a public good," namely whether the "secondary good" (what economists often describe in such contexts as an "externality") produced by small-scale booze makers—that is, a sense of *place*—is nonexcludable and/or nonrival. In other words, the authors consider whether the place-making effect of fermentation-focused development constitutes a public good in the economic sense of that term. Chapter 3, by John Overton, steps beyond the typical valorization of (developing) wine countries to examine those wine regions that do not succeed. Using several case studies in New Zealand to demonstrate his point, Overton describes a "landscape of failure" vis-à-vis the development of proximate, successful wine regions.

Part 2 addresses landscapes of ferment and includes chapters on the topics of wine, beer, spirits, cider, chocolate, and kombucha. In chapter 4 Christopher Holtkamp, Brendan Lavy, and Russell Weaver examine the varying production and consumption landscapes of Kentucky bourbon by investigating the history of the place and the product and evaluating marketing trends in the past and present. In chapter 5 Walter Furness and I take a look at how apples used for fermented apple cider ("hard cider") in Herefordshire, England, can be considered actors in the formation (and maintenance) of place and/or landscape using the lens of actor-network theory. In chapter 6 Mark Patterson, Nancy Hoalst-Pullen, and Sam Batzli discuss the various waves of migration to the United States and how those incoming cultural groups have influenced beer ingredients, brewing processes, and consumer style preferences. In chapter 7 Paul Zunkel describes the *goût du terroir* of Bloody Mary cocktails in the United States, demonstrating how regional cultural variations have conspired to make the drink unique from place to place while conserving the underlying elements of the "classic hangover" beverage. In chapter 8 Ryan Galt outlines some of the opportunities and challenges related to craft chocolate production in Hawai'i, especially as related to (subjective) notions of quality and the realities of production and marketing vis-à-vis tourism in the island state. Chapter 9, by Elizabeth Yarbrough, myself, and Colton Coiner, presents a case study of home-brewing "fermentos" to examine the wider cultural and community impacts of such practices via an ethnographic account of kombucha makers in San Marcos, Texas.

Part 3 takes a more meta-level view of the book's premise and includes chapters from both geographers and nongeographers seeking to explore how and why the metaphor of fermentation could provide insight in a variety of contexts. Chapter 10, by Maya Hey, traces the terrains of knowledge evident when one puts the same material—in this case cheese and cheese making—in two different spaces that both define and use that material differently. In chapter 11 Andy Murray dives deep into the meaning of zymurgeography (the geography of ferment), exploring the contours of and intersection(s) between scientific inquiry and fermentation. Chapter 12, by Eric Sarmiento, reflects on the microbiopolitical ecology of fermentation and considers how the microbiome and our relation to it help to make us more (engaged) human (actors). And in the penultimate chapter of the part and book, chapter 13, Vaughn Bryan Baltzly philosophizes on the (potential) utility of the concept of landscape ferment, examining the criteria by which such a conceptualization might be judged and in what circumstances it adds value to conversations around landscape change versus when it could be considered mere "spandrel" research. In chapter 14 I and my coauthors Walter Furness and Shadi Maleki offer some concluding words and invite further research.

These chapters, taken together, explore the idea and potential of fermented landscapes. While some offer conclusions about environmental and social processes and places in flux, others are more speculative, asking where we might go from here using the concept of fermentation as a paradigm for understanding place-based change.

This book thus takes an expansive, exploratory look at the concept of fermentation, applying it as a metaphor for change while revealing fermentation itself as driver of that change. By investigating the environmental, economic, and sociocultural implications of fermentation in both expected and unexpected places and ways, these cases home in on complex exchanges over time and space—an increasingly relevant endeavor in socially and environmentally challenged contexts, whether global or local.

Fermented Landscapes

Part 1

Conceptualizing the Role of
Fermentation in Processes of
Landscape Change

Fermented Landscapes

1

Considering the Macro Consequences of Micro(be) Processes of Socio-environmental Transformation

Colleen C. Myles

In this chapter I formulate and define the concept of fermented landscapes. I also offer some perspectives on fermentation, on (re)enlivening the contemporary food landscape (including some thoughts on human-microbe interactions), and on landscapes themselves. I provide a description of landscapes of ferment, or "fermented landscapes," and an explanatory case study, which leads into a discussion of the potential for a political ecology of fermented landscapes. This chapter thus serves as the launching point for the rest of the substantive chapters of the book.

On Fermentation

The fermentation of various food products is a long-standing strategy of food preservation and storage; humans have been using fermentation to produce and preserve nutrition and calories since at least the Neolithic Age (Fruton 2006). Fermentation—a process of biochemical change catalyzed by agents or (micro)organisms that cause organic substances to break down into simpler substances—is used in the creation or preservation of a wide variety of foods and drinks, including wine, beer, cider, spirits, kombucha, pickled cucumbers, kimchi, cheese, yogurt, chocolate, and more (Buchholz and Collins 2013; Pollan 2013). Cultural groups across the globe utilize a variety of methods for preserving food, including preventing the growth of bacteria, fungi (such as yeasts), or other microorganisms or introducing benign bacteria or fungi to the product. The fermentation of fruits and grains has historically been a means to assure food safety in times of questionable sanitation and environmental quality (Pollan 2013; Wood 2013). However, this food safety strategy proved to have admittedly amusing side effects because some of these endeavors—whether purposeful or not—resulted in the production of alcohol (Phillips 2014). Nevertheless, in contemporary times fermentation as a means of food preparation, safety, and preservation continues, and the recreational use of fermented (alcoholic and nonalcoholic) products still plays an important role in the cultural practices and foodways of many diverse cultural groups.

According to Michael Pollan (2013, 303), "If there is a culture that does not practice some fermentation of food or drink, anthropologists have yet to discover it. Fermentation would appear to be a cultural universal and remains one of the most important ways that food is processed." Indeed fermentation is a process that has been manipulated by humans for many thousands of years. Joseph Fruton (2006, xii) states that "human knowledge of the phenomena of fermentation is at least as old as agriculture," and it is likely much older (Pollan 2013). As Rod Phillips (2014, 6) notes, "We can trace alcoholic beverages made by humans to about 7000 BC, nine millennia ago, but it is almost certain that prehistoric humans consumed alcohol in fruits and berries many thousands of years earlier than that." Accordingly, Klaus Buchholz and John Collins (2013) call fermentation an "ancient handicraft"—even if it was a craft that humans did not fully understand: "That is not to say that our remote ancestors did not see such things as the mycelium that developed as a mould [sic] overgrew a food item, or the yeast mass that developed as a wine or beer underwent fermentation. However, it is far less certain that they understood that these were living things, indeed . . . arguments over (for example) yeast's nature and its mode of action, were still matters for heated debate until well into the 19th century" (Wood 2013, 2). Through scientific exploration and discovery, however, "mystical" theories of fermentation were rejected, though early conceptions of the process—including the idea that fermentation involves "several compounds act[ing] in harmony with each other" to first attract and then reject one another (Buchholz and Collins 2013, 3748)—survived. That said, fermentation is now firmly understood as a biological process (i.e., one intimately tied to *life*), which generates its own energy from within, acting like a "cold fire."[1]

Because the first human settlements preceded the invention of writing, "we cannot know how humans came to realise that some of the same agents that destroy foods can also act to conserve and improve them" (Wood 2013, 1–2), but somehow prehistoric humans did—and their descendants continue to do so to this day. There are several practical drivers for this practice (e.g., food safety, medical uses, cultural importance) and, not unimportantly, taste. As Pollan (2013, 294) puts it, "deliciousness is the by-product of decay." Heather Paxson (2008) goes further, noting that in the post-Pasteurian ethos, microbes are not just tasty but also ubiquitous and necessary, given our expanded notions of human health and nutrition (see also chapter 12); microbial life forms—bacteria, yeast, and mold—"not only contribute a kind of labor to the production of cheese and other fermented foods but also confer vitality on them" (Paxson 2008, 38). Bronwen Percival and Francis Percival (2017, 16) note the capacity of ferments like cheese to "link the biodiversity" of "three different worlds: flora, fauna, and microbiota"—and that they do so "in a form that the consumer can *taste*" (original emphasis). Indeed "many of the foods we eat are of microbial origin or contain constituents produced by microbial fermentation" (Seviour et al. 2013, 97), and innovations in food bioprocesses have had significant impacts on landscape and society. Microorganisms are increasingly seen as vital actors at the intersection(s) of cultural tradition, food production and politics, and (agrarian) landscapes (Paxson 2008; chapter 12, this volume).

Recent recognition of the depth of human-microbe interactions aside, it is clear that "humans have unwittingly and largely empirically been exploiting microbes for the production of foods for thousands of years and thus the food biotechnology industry is not a recent invention of western science" (Seviour et al. 2013, 97–98). Still, Louis Pasteur's introduction of "pure strains of yeast . . . transformed the making of wine and beer from a speculative enterprise, whose outcome was uncertain, into a science-based activity" (Fruton 2006, xiii).[2] It was the increasing demand for a range of fermented foods and products (e.g., vinegar, penicillin, and biodegradable polymers, not to mention wine, beer, and ethanol) that drove innovation in industrial-scale fermentation mechanisms and production methods (Wood 2013). By the end of the nineteenth century fermentation industries were growing rapidly in Europe, North America, and Asia.

Growth in fermentation industries (perhaps marking the start of the "Pasteurian era"?) had several drivers, including improvements in science and technology; the increasing economic importance of fermented products, which "encompassed the manufacture of beer and wine, industrial alcohol, yeast, acetic and lactic acid, cheese, soy sauce and sake" (Buchholz and Collins 2013, 3750); and growth of the perceived benefits of consuming fermented foods. For example, the purported physiological benefits of wine included "stimulation of the nervous system and blood circulation, improving or enhancing subjective feeling and performance" (Buchholz and Collins 2013, 3751). The prominence of alcohol in society is hardly new.

The Most Famous Ferment: Alcohol

"Of all humankind's fermentations," Pollan (2013, 374) writes, "alcohol is the oldest and by far the most popular, consumed in all but a small handful of cultures for all of recorded history and no doubt for a long time before that." Alcoholic beverages provide intoxication (which can be both pleasurable and valuable for medicine, religion, and art), a reliably safe drink (excepting overconsumption, which can cause a variety of ills for human health and society), a mode of effective calorie/food preservation, improved nutritional content, and a more complex taste.[3] Indeed some anthropologists think it was "the human desire for a steady supply of alcohol, not food, that drove the shift from hunting and gathering to agriculture and settlement" (385). Whether alcohol fermentation or agriculture came first, fermented food and drink served the role of diversifying the diet (including both taste and nutritional components) of newly settled humans as they undertook agrarian lifestyles and reduced the variety of foods consumed in comparison to a diet based on nomadic hunting and gathering.

While for some, alcohol is "associated with positive qualities like conviviality, fertility, and spirituality," others "have found history since the advent of alcohol resembling one long hangover for humanity" because alcohol has "negative associations such as social disruption, violence, crime, sin, immorality, physical and mental illness, and death" (Phillips 2014, 9). Accordingly, "an important dimension of the history of alcohol . . . is its contested status and the struggle to find a way to realize its benefits while minimizing its dangers" (3). However, even with alcohol's (potentially) negative associated outcomes, innovations in fermentation (such as those arising from the development of microbiology and biotechnology) have led to widely transformative social and environmental products that have revolutionized medicine, energy production, and environment management (Buchholz and Collins 2013).

(Re)enlivening the Contemporary Food Landscape

In contemporary times "the availability of pure cultures and our current understanding of microbial physiology and molecular biology have led to the development of highly controlled fermentation processes" (Seviour et al. 2013, 97–98), processes that have by design stripped fermented products of their liveliness while also assuring a more reliable yield and capacity to meet stringent food safety standards. While these food safety advances have been heralded as a great scientific success by some, it is precisely because of concerns over the sterilized state of many industrialized foods (even the fermented ones) that a movement for highly *un*controlled fermentation has emerged.[4] This is a movement that seeks to (re)enliven food and food culture. The standardization and regulation of fermentation processes—the Pasteurian approach— has changed the foods most people consume, making them less "lively" and (ironically) less safe because the microorganisms that protect fermented products arguably also protect us (Paxson 2008; chapter 12, this volume).[5]

People like fermentation guru Sandor Katz promote the consumption of "live-culture foods"—those that are "teeming with living bacteria"—for their health and wellness benefits (Pollan 2013, 297).[6] However, even Pollan admits that "the idea that the safety of a food is guaranteed by the bacteria still alive in it" (301) could be difficult for some people to accept. Nevertheless, and perhaps ironically, the list of foods that are in some way fermented is surprisingly long. It includes a number of highly common consumables, such as "coffee, chocolate, vanilla, bread, cheese, wine and beer, yogurt, ketchup and most other condiments, vinegar, soy sauce, miso, certain teas, corned beef and pastrami, prosciutto and salami" (304). However, "in many cases the role of fermentation in creating them is not widely understood" (304).

A Microbiology of Desire—or Human-Microbe Coevolution?

Because "fermented foods are typically both strongly flavored and strongly prized in their cultures," Pollan (2013, 304) suggests that "there may be a microbiology of desire at work in these foods, the bacteria and fungi having been selected over time for their ability to produce the flavors people find most compelling." In some instances (sourdough yeast, for example), certain microbial strains may exist nowhere else than in those specific ecological niches provided by humans (sourdough starters in kitchens worldwide), such that "one kind of culture uphold[s] the other" (304). Pollan suggests that, whether acknowledged or not, the "microcosmos," or the "unseen universe of microbes all around and within us" (297, referring to Lynn Margulis's term), is already an important part of cuisine, society more broadly, and even humanity itself.

As scientific (and lay) understandings of the body and its functioning have advanced, it has become clearer and clearer that humans are part of a web of life both inside and outside of the body (Buchholz and Collins 2013; Fruton 2006; McNeil et al. 2013). With this in mind, "fermentos" (people who ferment—and eat—all kinds of things at home [see chapter

9]), so-called "post-Pasteurians" (Pollan 2013), and others who follow the lead of (home) fermentation evangelists like Katz strive not only to consume more "live-culture foods" themselves but also encourage to others to adopt the same micro and macro life-changes. One aspect of these changes involves cultural revival via the sharing of materials (literal cultures) and ideas (social culture) within a community (see chapter 9 for a detailed discussion of "kombucha culture" in a particular locale). During these "culture swaps," fermentos are culturing (to use the biological term) two kinds of "culture": (human) tradition and practice, as well as microbiota (Pollan 2013).

While Pollan (2013) debates whether "coevolution" is too strong a term (mainly as it requires both actors to be changed), he ultimately argues that the term is appropriate. He describes how *Saccharomyces cerevisiae*—the yeast that serves as the catalyst and driver for the fermentation of beer and wine—has evolved and adapted to profit from ecological niches presented by humans; he then describes how humans have evolved (though some groups are still less capable of this) to metabolize and detoxify ethyl alcohol, making the fruits of *S. cerevisiae*'s labor palatable and useful. Beyond these metabolic and caloric benefits (which are exclusive to *S. cerevisiae*), *Homo sapiens* has also benefited from interspecies collaboration via the microbe's contributions to human spirituality and art. But what else has resulted from this interspecies collaboration or coevolution?

On Landscape(s)

For better or for worse, reconfiguring the earth has been humanity's prerogative, so much so that as human development has progressed from nomadic to agrarian to industrial, our impact on the planet has grown exponentially. Due to human intervention we have now entered a new geologic epoch, the Anthropocene (Whitehead 2014). Humanity's influence on natural systems has occurred at scales both large and small; human impacts on and impressions in the landscape are no exception. From this perspective, when considering how the production and consumption of ferments have made their mark on humanity and the earth, the most pertinent issues of interest are the attendant *landscape* impacts that result. As Pollan (2013, 384) notes, "The human desire for alcohol has been a tremendous boon to *Saccharomyces cerevisiae*. To supply it with endless rivers of liquid substrate to ferment, we have reconfigured vast swaths of the earth's surface, planting tens of millions of acres of grain and fruit, in the process creating a paradise of fermentable sugars to sustain this supremely enterprising family of fungi." The chapters in this volume examine the relationship between the production of fermentable things and the attendant landscape changes that result.

In light of geography's long-standing tradition of investigating the nature/society interface (Castree 2005), I see the concept of landscape as multifaceted (Mitchell 2000; Sauer 1963). The concept carries connotations of biophysical, environmental, and material realities and processes, as well as immaterial, symbolic, socially constructed realities and processes. Indeed the concept of landscape could be framed as *meaning-model-metaphor* (Hiner 2016a).[7] In this framework the meaning—in its technical definition (take *Merriam-Webster's* versions: "an area of land that has a particular quality or appearance" or "the landforms of a region in the aggregate")—has particular modes and models (like landscapes of production, industry, or urbanization) and also serves as a metaphor, carrying connotations beyond its strict (physical, observable) manifestations.

Thus, when I refer to a "landscape of fermentation" in these pages, I am invoking both the physical, ecological elements related to the production and consumption of fermented products as well as the symbolic, cultural, and sociopolitical ramifications of that landscape form (Mitchell 2008).

A landscape model that is especially relevant to this research program is the concept of the rural-urban interface, or the place where rural and urban meet, mingle, and mix (Hiner 2014; Lichter and Brown 2011). Some argue that there is no real separation between so-called "rural" and "urban" spaces, especially in an era of global urbanization (McCarthy 2008; Scott, Gilbert, and Gelan 2007; Woods 2007). However, the distinction between the urban and rural can have real meaning for stakeholders in each setting—and for those in settings situated between them (Abrams et al. 2012; Heley and Jones 2012; Hiner 2015, 2016b; Travis 2007). Specifically, I and others have argued that physical landscapes vary along the interface, and, importantly, so do sociocultural ones. The meanings and values attributed to particular places matter; how land managers, local residents, and policy makers at a variety of scales perceive the environmental context of a place makes a difference in terms of how they manage it (Cadieux and Taylor 2013; Hiner 2016c; Taylor and Hurley 2016; Walker and Fortmann 2003; Walker and Hurley 2011). Moreover, the linkages, exchanges, and one might even say metabolisms between the city and country—both material and symbolic—are also related to how people view and interact with rural versus urban places (McKinnon et al. 2017).

Fermented Landscapes Defined

One version of this rural-urban linkage is exemplified by the concept of a fermented landscape, a framework that examines the production, distribution, and consumption of fermented products (such as alcoholic beverages and beyond) as a focal point in the study of complex rural-urban exchanges or metabolisms over time and space. In this way the fermented landscapes framework provides a lens through which to uncover ongoing processes of change and transformation in both expected and unexpected places and ways. The research presented here explores the transformation—whether radical, slow, or something in between—of landscapes that results from the "fermentation" of place, namely the active, sometimes volatile changes catalyzed by the mixing of some (f)actor with another.

Biochemical metabolic fermentation is marked by the colonization of a fermentation substrate by microbes, but the success of that fermentation—that is, who or what becomes the dominant occupier of the space—is influenced by "temperature, place, and chance" (Pollan 2013, 316; see also Paxson 2008). When used metaphorically or figuratively, the same logic stands: which actors (human or not) are dominant as a place is transformed will depend on several factors, including the context (temperature), the location (place), and sometimes simple chance. In other words, the environmental, economic, sociocultural, and geographic elements of a place, paired with random luck, will determine the context within which the change will occur. Fermentation constitutes a particular kind of change; it is both microscopic and wholly encompassing, and it includes a succession of actors and biochemical or material transformations. Paxson (2008, 25) asks, "Is there a role for microbes—and microbiopolitics—in thinking through links between land and food, place and taste?" She argues that for post-Pasteurians artisanal fermented products are essential to the sustainability and success of working landscapes by serving as key avenues for the expression of localism, maintaining connections to the land, and preserving shared lifeways. Taking it further, I ask—and the authors in this volume explore—the wider landscape effects of microbial, material ferment and of more metaphorical fermentations: How might landscapes be transformed by or through the process(es) of fermentation?

This book offers numerous cases that explore the potential of fermented (i.e., catalyzed, creative, leavened) landscape change (either separate from or in contrast to "regular" landscape change [see chapter 13, this volume]). These stories emerge from a collection of work undertaken individually and in collaboration with others. As described within individual chapters, methods for this research include ethnography, surveys, content analysis, reviews of secondary economic and sociodemographic data, and more.

A Case in Point: Wine in the Sierra Nevada

My curiosity about the land management and cultural implications of fermentation-focused development began in the scenic Sierra Nevada foothills of California. In many ways marked by its past, it is a landscape whose previous forms and uses have been both boon and bane (Duane 1999). Its history of primary production, especially the extraction of minerals and raw materials for use elsewhere, in urban places near and far, has left marks on the landscape. As a place that has weathered several boom and bust economies, the Sierra Nevada is home to a wine industry that is in some respects just the latest in a string of environmentally extractive businesses that cater to and serve outside interests (Myles and Filan 2017). But it is more than just the extraction; the shifts are not just environmental but also cultural. While wine grape production and winemaking are not new to the Sierra Nevada, the industry has become increasingly prominent economically and culturally, as evidenced by the visible elements of the trade (e.g., vineyard development, wine tasting infrastructure, the number and size of wineries present) and the associated increase in wine-based tourism and amenity-focused in migration (Myles and Filan 2019)—all of which involve landscape change. The economic development afforded to the region via these changes is clear; however, the beneficiaries (and victims) of such transformations are not always as obvious (see chapter 2 for more on this and chapter 3 for an exploration of regions that fail to achieve economic transformation).

For an illustration of this point, consider figure 1, a photo captured on a wine trail in busy Amador County wine country in the Sierra Nevada foothills of California. Pictured is a landscape that is manicured, prepared, cultivated—one that is decidedly different than a landscape focused solely on production. In short, this landscape was made to be consumed. This tidy vineyard presents a clean, approachable landscape for visitors (many of whom travel from nearby urban centers); this landscape quality is just as important for the success of the wine region as the taste or quality of the wine produced. Just around the bend in the road a slew of tasting rooms sit ready for consumers and poised to sell not just the tasty drop produced on these lands but also the image of the land and lifestyle itself. And, as this image shows, the management of the land reflects its multiple purposes, as a place of production (e.g., the tractor-crossing sign) but also of consumption (e.g., the well-kept fence and ornamental plants).

FIG. I. "Cultivated to be consumed": a constructed image of wine country in Amador County, California, located in the Sierra Nevada foothills. Photo by Colleen C. Myles, July 2014.

In the Sierra Nevada, as agricultural producers, craftspeople in the fermentation business, and entrepreneurs and investors endeavor to make a place for themselves in a competitive and increasingly global marketplace, wine is a niche carved out to mobilize multiple beneficial ends—to promote economic development while also preserving the open landscapes and environmental quality of the place (Myles and Filan 2019). In this way wine grapes and wine are a part of an ongoing process of not just environmental change but of cultural transformation, a transformation that is intimately related to the flux and exchange of rural and urban people, material, and ideas. In terms of ferment, this landscape has shifted from one form to another through the influx of new actors and their spontaneous inter-reactions to create a place different from what it was before. In other words, it now stands as a fermented landscape.

Paysages de la fermentation

The study of wine grape growing and wine production in the Sierra Nevada led to my deep desire to try to understand the form, function, and meaning of landscapes of ferment not only in their constituent parts—environment, economy, culture—but also as comprehensive wholes. As such, the fermented landscapes approach pursues just such a holistic analysis of place, seeking to understand its relation with other places and its significance to the people within it. Specifically, the fermented landscapes approach delves into the how and why of various (physical and cultural) landscapes vis-à-vis literal and figurative processes of fermentation. When I described the concept to a French colleague, he recommended translating it as *paysages de la fermentation*, rather than a literal translation of *paysage fermenté*, because "fermented landscapes" suggests an image of the land itself bubbling up in ferment.[8] While I agree that *paysages de la fermentation* (landscapes of fermentation) is more fitting as an allusion to the environmental and sociocultural elements of fermentation that this kind of research approaches, the image of a landscape of active ferment is a telling one, and not too far off base, depending on the context.

Food, Fermentation, and Place-Making:
A Political Ecology of Fermented Landscapes?

For "fermentos," fermentation is "much more than a way to prepare and preserve food"; it is "a political and ecological act, a way to engage with the bacteria and fungi, honor our coevolutionary interdependence, and get over our self-destructive germophobia" (Pollan 2013, 297). Citing the "wars on bugs" spurred by early microbiological research, people in the post-Pasteurian camp argue that we should "rehabilitate the image of bacteria," fight back against the sterilization, cultural homogenization, and industrialization of our food, "break the dependency of consumerism, [and] rebuild local food systems (since fermented foods allow us to eat locally all year long)" (299–300). A commitment to fermentation can thus be both destructive to the status quo and also ultimately creative and enlivening. As Pollan notes, "Though we've tended to think of bacteria as agents of destruction, they are, like other fermenters, invaluable creators as well" (327).

Similarly, in each of the cases described in this book fermentation plays a crucial role in the politics of place-making and associated processes of ecological change. What these stories tell us is that there can be clear economic and social benefits to communities as they transition into fermentation-focused development, but there are also detriments. In other words, although pursuing economies based on production and consumption of food and ferments is increasingly popular, there is much to do to create a just, diverse, local food system. By calling attention to the actual (both positive and negative) impacts of the fermentation-focused, local food movement on landscapes and communities, we may better recognize the processes of power shaping the food and fermented landscapes that are emerging. And if we see political ecology as seeking to understand how power and privilege mediate human interactions with the environment (Robbins 2011), the cases described in this volume show us how an analytical framework based on the concept of fermented landscapes can be applied to a variety of sites of human-environment interaction.

Research within the fermented landscapes framework examines the excitement, unrest, and agitation evident across shifting physical-environmental and sociocultural landscapes, especially as related to the production, distribution, and consumption of fermented products. In this way the "fermented" in fermented landscapes has a double meaning, referring to both the literal fermentation of a given product as well as the figurative fermentation of the place (the term "place" encapsulating both environment and culture). Thus, a study of a fermented landscape is a (re) consideration of both material and semiotic (aka symbolic) change in a given context. As such, research falling under this conceptual umbrella might ask questions like: What can we learn from landscapes dominated by fermentation? How are they distinct from other forms of landscape change or (re)development? Who or what is driving the changes taking place? Relevant research could include wine, beer, and cider geographies (and the geography of other fermented products, whether alcoholic or not), as well as more philosophical or theoretical examinations of the power of the fermentation metaphor in science and society—and this volume offers examples of each.

Notes

1. This particular view of fermentation led early investigators such as Pasteur to refer to fermentation as a "boil," even though "this particular boil wasn't hot to the touch" (Pollan 2013, 295).

2. Furthermore, "the understanding that microbes exist, obey biological laws and can be used in a controlled way, are essentially outcomes of our development of the scientific method" (Wood 2013, 2). Indeed "Pasteur's work led to the establishment of the science of microbiology," which at its inception was rooted in "fascinating discoveries, such as yeast as living matter being responsible for the fermentation of beer and wine" (Buchholz and Collins 2013, 3747).

3. With regard to medicine, Pollan (2013, 382) states that "alcohol is a powerful and versatile drug" that "for most of human history was the most important drug in the pharmacopeia" due to its stress- and pain-relieving capabilities (see also Phillips 2014).

4. According to Pollan (2013, 332–33), "The so-called Western diet, with its refined carbohydrates, highly processed foods, and dearth of fresh vegetables, is downright hostile to fermentation: It preserves foods by killing bacteria rather than cultivating them, and then deprives our gut bacteria of much of anything good for it to ferment."

5. And Pasteurian standardization and regulation is truly a politics of control: "Pasteurianism is a biopolitics predicated on the indirect control of human bodies through direct control over microbial bodies" (Paxson 2008, 36).

6. However, because active ferments reside in a place that is just this side of funky, such foods are often an acquired taste, thus supporting the idea that the concept of "rotten" is culturally constructed. For example, Pollan (2013, 303) tells the story of eating decomposing shark and, though it is a local delicacy in Iceland, finding it barely palatable.

7. The use of fermentation as a lens or frame through which to view places and processes builds on previous work dealing with the application and expansion of terms as a means through which change is made. Using the example of the rural-urban interface, I present the concept as simultaneously containing a meaning, serving as a model, and emerging as a metaphor for places in flux (Hiner 2016a). This

progression from meaning (a technical, descriptive definition), to model (a larger framing model of interactions), to metaphor (a heuristic device that can be applied to greater concepts, issues, and places in order to glean greater understanding or analytical purchase) explains conceptual deepening over time.

8. I am grateful to Dr. Yvon Le Caro for offering this helpful insight.

References

Abrams, Jesse, Hannah Gosnell, Nicholas Gill, and Peter Klepeis. 2012. "Re-Creating the Rural, Reconstructing Nature: An International Literature Review of the Environmental Implications of Amenity Migration." *Conservation and Society* 10 (3): 270–84.

Buchholz, Klaus, and John Collins. 2013. "The Roots—A Short History of Industrial Microbiology and Biotechnology." *Applied Microbiology & Biotechnology* 97 (May): 3747–62.

Cadieux, Kirsten, and Laura Taylor. 2013. *Landscape and the Ideology of Nature in Exurbia: Green Sprawl.* New York: Routledge.

Castree, Noel. 2005. *Nature.* New York: Routledge.

Duane, Timothy P. 1999. *Shaping the Sierra: Nature, Culture, and Conflict in the Changing West.* Berkeley: University of California Press.

Fruton, Joseph S. 2006. *Fermentation: Vital or Chemical Process?* Boston: Brill.

Heley, Jesse, and Laura Jones. 2012. "Relational Rurals: Some Thoughts on Relating Things and Theory in Rural Studies." *Journal of Rural Studies* 28 (3): 208–17.

Hiner, Colleen C. 2014. "'Been-Heres vs. Come-Heres' and Other Identities and Ideologies along the Rural-Urban Interface: A Comparative Case Study in Calaveras County, California." *Land Use Policy* 41 (November): 70–83.

———. 2015. "(False) Dichotomies, Political Ideologies, and Preferences for Environmental Management along the Rural-Urban Interface in Calaveras County, California." *Journal of Applied Geography* 65 (December): 13–27.

———. 2016a. "Beyond the Edge and in Between: (Re)conceptualizing the Rural-Urban Interface as Meaning-Model-Metaphor." *Professional Geographer* 68 (4): 520–32.

———. 2016b. "'Chicken Wars,' Water Fights, and Other Contested Ecologies along the Rural-Urban Interface in California's Sierra Nevada Foothills." *Journal of Political Ecology* 23 (1): 167–81.

———. 2016c. "Divergent Perspectives and Contested Ecologies: Challenges in Environmental Management along the Rural-Urban Interface." In *A Comparative Political Ecology of Exurbia: Planning, Environmental Management, and Landscape*

Change, edited by Laura Taylor and Patrick T. Hurley, 51–82. Cham, Switzerland: Springer.

Lichter, Daniel T., and David L. Brown. 2011. "Rural America in an Urban Society: Changing Spatial and Social Boundaries." *Annual Review of Sociology* 37:565–92.

McCarthy, James. 2008. "Rural Geography: Globalizing the Countryside." *Progress in Human Geography* 32 (1): 129–37.

McKinnon, Innisfree, Patrick T. Hurley, Colleen C. Myles, Megan Maccaroni, and Trina Filan. 2017. "Uneven Urban Metabolisms: Toward an Integrative (Ex)urban Political Ecology of Sustainability in and around the City." *Urban Geography.* https://doi.org/10.1080/02723638.2017.1388733.

McNeil, Brian, David Archer, Ioannis Giavasis, and Linda Harvey, eds. 2013. *Microbial Production of Food Ingredients, Enzymes and Nutraceuticals.* Cambridge: Elsevier Science & Technology.

Mitchell, Don. 2000. *Cultural Geography: A Critical Introduction.* Malden MA: Blackwell.

———. 2008. "New Axioms for Reading the Landscape: Paying Attention to Political Economy and Social Justice." In *Political Economies of Landscape Change: Places of Integrative Power*, edited by James L. Wescoat and Douglas M. Johnston, 29–50. New York: Springer.

Myles, Colleen C., and Trina Filan. 2017. "Boom-and-Bust: (Hi)stories of Landscape Production and Consumption in California's Sierra Nevada Foothills." *Polymath* 7 (2): 76–89.

———. 2019. "Making (a) Place: Wine, Society, and Environment in California's Sierra Nevada Foothills." *Regional Studies, Regional Science* 6 (1): 157–67.

Paxson, Heather. 2008. "Post-Pasteurian Cultures: The Microbiopolitics of Raw-Milk Cheese in the United States." *Cultural Anthropology* 23 (February): 15–47.

Percival, Bronwen, and Francis Percival. 2017. *Reinventing the Wheel: Milk, Microbes, and the Fight for Real Cheese.* Oakland: University of California Press.

Phillips, Rod. 2014. *Alcohol: A History.* Chapel Hill: University of North Carolina Press.

Pollan, Michael. 2013. *Cooked: A Natural History of Transformation.* New York: Penguin.

Robbins, Paul. 2011. *Political Ecology: A Critical Introduction.* Malden MA: Wiley-Blackwell.

Sauer, Carl Ortwin. 1963. "The Morphology of Landscape (1925)." In *Land and Life: A Selection from the Writings of Carl Ortwin Sauer*, edited by John Leighly, 315–50. Berkeley: University of California Press.

Scott, Alister, Alana Gilbert, and Ayele Gelan. 2007. *The Urban-Rural Divide: Myth or Reality?* SERG Policy Brief No. 2. Aberdeen, Scotland: Macaulay Institute.

Seviour, Robert J., Linda M. Harvey, Mariana Fazenda, and Brian McNeil. 2013. "Production of Foods and Food Components by Microbial Fermentation: An Introduction." In *Microbial Production of Food Ingredients, Enzymes and Nutraceuticals*, edited by Brian McNeil, David Archer, Ioannis Giavasis, and Linda Harvey, 97–124. Cambridge: Elsevier Science & Technology.

Taylor, Laura, and Patrick T. Hurley, eds. 2016. *A Comparative Political Ecology of Exurbia: Planning, Environmental Management, and Landscape Change*. Cham, Switzerland: Springer.

Travis, William R. 2007. *New Geographies of the American West: Land Use and the Changing Patterns of Place*. Washington DC: Island Press.

Walker, Peter, and Louise Fortmann. 2003. "Whose Landscape? A Political Ecology of the 'Exurban' Sierra." *Cultural Geographies* 10 (4): 469–91.

Walker, Peter A., and Patrick T. Hurley. 2011. *Planning Paradise: Politics and Visioning of Land Use in Oregon*. Tucson: University of Arizona Press.

Whitehead, Mark. 2014. *Environmental Transformations: A Geography of the Anthropocene*. New York: Routledge.

Wood, Brian J. B. 2013. "Bioprocessing as a Route to Food Ingredients: An Introduction." In *Microbial Production of Food Ingredients, Enzymes and Nutraceuticals*, edited by Brian McNeil, David Archer, Ioannis Giavasis, and Linda Harvey, 1–15. Cambridge: Elsevier Science & Technology.

Woods, Michael. 2007. "Engaging the Global Countryside: Globalization, Hybridity and the Reconstitution of Rural Place." *Progress in Human Geography* 31 (4): 485–507.

Booze as a Public Good?   2

How Localized, Craft Fermentation Industries
Make Place, for Better or Worse

*Colleen C. Myles, Christopher R. Holtkamp, Innisfree McKinnon,
Vaughn Bryan Baltzly, and Colton Coiner*

Increasing awareness of and engagement with locally produced craft products has been a hallmark of the twenty-first century, as consumers actively seek out businesses that offer distinct connections to place and/ or locally produced goods and services (Donald 2009).[1] This interest in local products has fostered an explosion of commercial craft producers of food and fermented beverages in particular, with small-scale wineries, breweries, and distilleries proliferating in both rural and urban environments (Stopa 2016; Shoup 2017).[2] Businesses not only benefit from this interest in place but are active agents in the production of place across rural and urban landscapes.

In this chapter we explore what it means to conceptualize "booze" — wine, beer, cider, and spirits — as a public good. We inquire whether various place-making externalities associated with craft fermentation can be framed as something that is nonexcludable and nonrivalrous (i.e., a public good), regardless of whether those externalities are seen as normatively good or bad. A dictionary defines a "good" in the economic sense as a commodity or service that can be used to satisfy human wants and that has exchange value. Economic goods come in several forms: private goods, common pool resources, club goods, and public goods (Beggs 2018). Public goods are generally characterized as being nonexcludable and nonrivalrous, which means first that the benefit or impact of the good cannot be withheld from anyone who might want it or who might come into its orbit and second that one person's enjoyment or experience of the good does not diminish another person's enjoyment or experience of it (Beggs 2018; Cornes and Sandler 1996). Through a series of ethnographic case studies across several U.S. research sites, we examine how craft producers create "dreamscapes of visual consumption" (Zukin 1991, 221) or, as conceptualized in chapter 1, fermented landscapes. We examine how in these cases various actors alter the physical environment and shape sociocultural landscapes to determine whether and, if so, how craft producers have effected nonexcludable and nonrivalrous changes (both positive and negative) in their communities.

Literature Review

Changes in Consumer Behavior

According to Chris Holtkamp et al. (2016, 66), "Progress in technology led to the mass production and transport of consumables, helping to create an indistinguishable, flattened geographic . . . landscape." This homogenization of production and consumption led to a loss of identity and sense of place. More recently, however, consumers have begun to rebel against this effect (Flack 1997; Schnell and Reese 2003; Zelinsky 2011). Growing cultural awareness, environmental considerations, concern over rural decline, and concerns about quality are pushing consumers to select locally produced goods. Betsy Donald (2009, 2) states that "for the quality-seeking consumer . . . , it may be about consumer products grown locally; for another it may be about buying products free from certain allergens, synthetic additives, pesticides or herbicides regardless of the source." There has been tremendous growth in the presence of so-called "alternative" food systems and networks in the past several decades, reflecting an overall increase in awareness of and preference for locally produced and consumed goods (Maye, Holloway, and Kneafsey 2007).

Consumers are searching for individualism, shifting away from mass production and toward products that reflect local character and differentiation (Dawson and Burt 1998). Particularly with fermented products like beer, wine, and spirits, consumers deliberately choose products that produce or reinforce their identity (SIRC 2000; Schnell and Reese 2014). For example, a "2015 Nielsen study found that 52 percent of all craft beer consumers (55 percent of those aged twenty-one to thirty-four) consider whether a beer is local in their purchasing decision" (Holtkamp et al. 2016, 68). Wineries, breweries, and distilleries (among others) are responding to consumer demand for place connection by adopting local place-names and using locally sourced agricultural products (Schnell and Reese 2014; Holtkamp et al. 2016; Minnick 2016).

Consumers are also willing to pay higher prices for locally produced goods, partly because of a perception of higher quality (Atkins and Bowler 2001). Producers are responding by actively linking their goods to the place of origin as well as highlighting sourcing and production processes (Ilbery et al. 2005). And even when local sourcing is not possible, the act of craft production makes the products local; for example, in urban (industrial) places a kind of material transformation can occur through the process of brewing, wherein nonlocal, nonurban inputs become precisely that (Myles and Breen 2018). As consumers are turning against mass-produced, industrial-scale products, producers are embracing their small scale as an asset (Hindy and Hickenlooper 2015; Shears 2014). Craft brewers' small-scale production—whether or not said production is also "local" (the concepts of "small-scale" and "local" often being conflated)—is part of the attraction (Davis 2015).

Additional impetus for the explosion of craft producers comes from the rapidly changing regulatory environment (Garavaglia and Swinnen 2018). In the United States in places where alcohol production and distribution had been severely restricted, states have liberalized their laws in order to benefit from growing interest in craft fermentation (Thompson 2018). One significant change has been a cut to the federal excise tax on distilled spirits. An 80 percent decrease in the tax rate for craft producers making fewer than one hundred thousand gallons annually has been a windfall, leading many to invest in expanded marketing and distribution and others to drop the prices of their final product (Simonson 2018). This, combined with declining costs of production equipment and easier access to marketing, has contributed to the rise of craft producers, even as overall consumption has fallen. Indeed, even as large-scale commercial producers struggle with declining sales, craft production is expanding at their expense (K. Taylor 2016; Wagner and Metzger 2017; Thompson 2018). In 2017, as overall beer sales fell by approximately 1 percent, craft beer sales increased by 5 percent, reaching nearly 13 percent of the U.S. market (Brewers Association n.d.).

Rural Landscapes

Producers have responded to the interest in locally produced products by actively emphasizing and celebrating connections to their sites of production, a trend noticeable across the food system (Morgan, Marsden, Murdoch 2009). One way this is achieved is through the use of locally sourced raw materials. A recent study of bourbon producers (see chapter 4, this volume) found that nearly two-thirds (62 percent) utilized local grains; craft bourbon producers are deliberately seeking locally produced source materials from small farmers and using that practice as a sign of quality inherent in the finished product. By linking "individual production locales, particular production processes[,] or specific agricultural and craft products," producers leverage the perceived quality consumers associate with that relationship (Ilbery et al. 2005, 118). The focus on local products is affecting the rural landscape as farmers respond to increased demand by changing what they produce and how they produce it (Zasada 2010). In North Carolina and Virginia the expansion of the craft brewing industry has led area farmers to turn to hop production in order to meet the growing demand for locally sourced ingredients (Hayward and Battle 2017). The growing popularity of ciders is also altering rural landscapes: "[Ciders] produce a very particular kind of landscape, often with very high levels of biodiversity not only among the fruit trees themselves, but also the flora and fauna in the meadows and fields that house the trees. When people drink more of this cider, there are more of these landscapes. When tastes for the cider wane, trees get felled" (Jordan 2016, 6).

In this way fermented apple cider production is a driver of physical and cultural landscape form and function (see chapter 5, this volume). These impacts can be considered positive externalities created by burgeoning craft production as local farmers benefit from a new customer base and diversified crop portfolio and as expanding cider orchards contribute to increased biodiversity and habitat.

In addition to influencing agricultural production, craft fermentation can contribute to rural tourism, attracting visitors who want to experience not just a beverage but the act of producing that beverage as well (Kline, Slocum, Cavaliere 2017; Slocum, Kline, Cavaliere 2018). Telling the story of production adds value for the business by serving the tourists' desire for authenticity and connection and may contribute to economic sustainability in rural areas (Sims 2009).

Some rural landscapes have felt the impact of both booms and busts— and sometimes multiple cycles of each—as cultural and environmental landscapes have shifted over time (Myles and Filan 2017). Take, for example, the Sierra Nevada foothill region in California, which has experienced drastic environmental and cultural change related first to primary resource extraction (mining, timber, agriculture) and later to amenity-focused landscape transformation (Duane 1999; Walker and Fortmann 2003; Beebe and Wheeler 2012; Hiner 2014; Chase 2015). Most recently, such amenity-focused landscape change has been driven by the making (or better, the re-presenting) of a wine landscape (Myles and Filan 2019).

Rural and exurban regions like the Sierra Nevada, whether established or developing as an amenity destination, follow a similar pattern, as the economic, environmental, and cultural emphasis across the landscape shifts from production to consumption (L. Taylor and Hurley 2016). Amenity landscapes are produced through the slow or rapid, haphazard or methodical remaking of a place into a desirable destination for tourism, or for a second home, or for a new life for new residents to the area—perhaps commodifying the landscape but certainly altering its fundamental character, for better or for worse (Travis 2007; Gosnell and Abrams 2011; Larsen and Hutton 2011; L. Taylor 2011; Argent et al. 2013; Senese et al. 2018). This transition can have both positive and negative impacts on the physical and cultural landscape. In one regard, commodifying the landscape may result in more interest in preserving that landscape because of the value it provides in supporting consumption by tourists. Alternatively, some owners who choose not to participate in the tourism industry may be pressured to accommodate themselves to the new aesthetic of the region or potentially be forced out altogether due to increased land values and other challenges—in effect, rural gentrification.

Urban Landscapes

Craft production is also having a significant impact on urban landscapes (Zukin 1991). This is occurring primarily through producers acting as first-wave gentrifiers, moving into economically struggling or otherwise underdeveloped areas and contributing to their revitalization (Donald 2009; Mathews and Picton 2014; Hubbard 2016). As some scholars have noted, "craft breweries become the canary in the coal mine for neighborhood change" as they appear in economically depressed areas due to low rents, fewer barriers for permitting nuisances (noise, odors, wastewater pollution), and other related factors (Barajas, Boeing, and Wartell 2017, 4). Craft producers also attract complementary businesses seeking to serve the same customers and can draw investment to improve nearby amenities and facilities, public or private (Crawford 1992; Zukin 2011; Hubbard 2016; Barajas, Boeing, and Wartell 2017). The phenomenon wherein normally competitive businesses cooperate for mutual gain can be referred to as "coopetition" or cooperative economics (Myles and Breen 2018). Whatever we call the phenomenon, it can transpire when incoming entrepreneurs help develop place identity through installation of art spaces, boutiques, restaurants, microbreweries, and the like, all by drawing in investment and people and thus putting in place an economic multiplier effect. While often beneficial for those involved, these changes may not be uniformly perceived as positive, as discussed below.

Besides the benefits accruing from craft producers' direct contributions to neighborhood change, these businesses may also influence regulatory changes and public infrastructure investments that benefit other economic activities and the community as a whole (Perritt 2013; Hopkins 2014; Best 2015; Chapman, Lellock, and Lippard 2017). For example, in Oregon the City of Portland (n.d.) has developed specific wastewater discharge requirements for business owners within the craft fermented beverage industry, including "breweries, distilleries, wineries, cideries, meaderies, and kombucharies." These investments and associated activities contribute to the physical and cultural landscapes of these places—the key components of a "sense of place"—as a location becomes desirable because of the mix of businesses and attractions associated with that place (Zukin 2011; Patterson and Hoalst-Pullen 2014; Reid and Gatrell 2017; Gatrell, Reid, and Steiger 2018; Slocum, Kline, and Cavaliere 2018).

Thus, craft production has a significant impact on the area in which it locates. In some ways this impact is positive because it fosters economic growth and physical improvements to the neighborhoods where producers choose to locate. However, it is important to recognize the potential negative externalities, particularly that of gentrification and the creation (albeit unintentional) of exclusionary places where former residents may not feel welcome in the neighborhoods they long occupied.

Production of Place

Craft producers recognize the value of place attachment and seek to connect to place through the adoption of local place-names for their business and/or products. They benefit from the shared culture and history associated with connection to place (Tuan 1991). Consumers want to feel like they are connecting to a place; consuming locally produced goods reinforces that experience (Holtkamp et al. 2016). Consumption becomes an "expression of authenticity, real food, small capitalism, locality, working-class community and regionality" (Spracklen, Laurencic, and Kenyon 2013, 317). An example of connection to place can be found in Kentucky bourbon marketing. Many distillers make explicit connection to Kentucky history and early settlers who started distilling in the state through the name they choose for their business or by using names like Elijah Craig, Evan Williams, and Jim Beam for their products. This practice allows even new distilleries to make a connection to history by purchasing the naming rights to historic brands and distilleries no longer in business (McKeithan 2012). When consumers drink Kentucky bourbon, they feel like they are a part of the history and the unique character and identity associated with that area, reflecting the lasting impressions of heritage in place over time (Myles and Filan 2017), even if that connection is not as authentic as it may seem.

Another benefit that may come from connection to place is an increase in social capital. Social capital can be defined as "the sum of actual and potential resources embedded within, available through, and derived from the network of relationships possessed by an individual or social unit" (Nahapiet and Ghoshal 1998, 243). Communities with higher social capital value engagement and have the capacity to foster positive change because of the networks and relationships that are the foundation of social capital (Putnam 1993). Establishments designed for drinking, as well as the act of drinking, facilitate social interactions and bonding, which may lead to increased social capital as people from different classes and backgrounds come together over craft products (SIRC 2000; Campbell 2005). The networks and relationships forged over craft products become a positive externality as they contribute to higher social capital in neighborhoods and communities where these establishments are found.

Detriments of Craft Production

Despite the myriad positives associated with the rise of craft fermentation, it is important to recognize potential drawbacks. Two notable issues regarding craft producers are their role in gentrification and the lack of diversity associated with the business and consumption of craft (fermented) products. This presents a challenge, particularly in urban environments, where these producers are proliferating. Indeed, insofar as the "benefits" of such place change are inescapable—that is, insofar as every person inhabiting these transformed spaces is affected by these changes (these "benefits" being thus nonexcludable) and insofar as no inhabitant's receipt of these "benefits" erodes their impact on anyone else (these impacts being thus nonrivalrous)—such place change represents a public good in the technical sense of the term (Cornes and Sandler 1996; Beggs 2018). However, the creation of that public good may also generate negative externalities affecting other members of the community, especially those displaced by gentrification.

Gentrification generally refers to the process of middle-class professionals moving to disinvested central city neighborhoods, upgrading housing, and attracting new businesses that cater to the new neighborhood clientele (Lees, Slater, and Wyly 2008; Zukin 2011; Mathews and Picton 2014; Hubbard 2016). This process often leads to displacement of low-income and minority residents from neighborhoods they have often occupied for decades. The appeal of attracting craft producers to serve economic development and place-making interests often contributes to changes in the regulatory environment (Barajas, Boeing, and Wartell 2017; City of Portland n.d.), and sometimes incentives to attract these types of businesses come at the expense of established, though low-income, neighborhoods (Arbel 2013; Perritt 2013; Best 2015). Whereas craft producers are often associated with the creation of place, the place created, it seems, may primarily be built for upper-income white consumers, and those that do not fit that description are displaced in the transition.

Some scholars (Alkon 2012; Alkon and Agyeman 2011; Maye, Holloway, and Kneafsey 2007) have offered helpful discussions on the role of race and class in the construction and maintenance of alternative food systems, within which craft beverages are often situated (Chapman, Lellock, and Lippard 2017). Although there are indications that craft production is becoming more diverse, with growing numbers of participants of color, women, and young people (Watson 2014), the majority of craft beverage producers and consumers remain higher-income white persons (Campbell 2005; Infante 2015; Thurnell-Read 2016; Barajas, Boeing, and Wartell 2017). Jesus Barajas, Jeff Boeing, and Julie Wartell (2017, 11) noted that between 2000 and 2010 "breweries were more likely to locate in census tracts that lost racial and ethnic diversity." In fact, they continued, "for each percentage point decline in the black and Latino populations, breweries were about three percent more likely to open" (11). This loss of diversity relates to the issue of gentrification, as discussed above, because craft producers act as both instigators and beneficiaries of neighborhood change, and while craft fermentation industries can be *exclusive* in terms of who participates and who benefits, the community and economic impacts of such industries are not *excludable* because they are universally imposed.

In order to explore the changes, both positive and negative, associated with craft production, we present a series of case studies, in both rural and urban environments, that reflect the impacts craft fermentation has had on communities and regions of Wisconsin, California, Texas, and Kentucky.

Case Studies

The case studies included in this chapter all stem from embedded, qualitative methods, including (participant) observation, interviews, and focus groups, as well as analysis of secondary data (including public data and promotional materials), conducted over several years by the chapter authors.

Craft Production in Wisconsin's Driftless Area

Situated in southwestern Wisconsin, the Driftless area is largely rural, in contrast to the sprawling urban centers on the eastern side of the state. Because of the area's hilly, forested topography, the significant concentration of organic farms, and tendency to vote for Democrats, the Driftless has been described as being more like Vermont than the Midwest (Nixon 2000). Tourism in the region is largely centered on the many opportunities for trout fishing (Orrick 2016), but other attractions include scenic drives along the upper Mississippi, the Amish farms where one can buy handcrafted goods and hanging baskets overflowing with flowers, and the growing local food, beer, and wine scene (Driftless Wisconsin n.d.).

Local beer, wine, and cider fits well into the region's focus on natural beauty, picturesque small farms, and a growing local food scene (Driftless Wisconsin n.d.). While many small farms transport goods to markets in Madison or Minneapolis or distribute through food hubs or co-ops, others focus on attracting consumers into the picturesque countryside, where they consume not only food, lodging, and recreation but also the landscape more broadly (McKinnon 2016). Isolated rural bars are a common sight among the scenic hills (Flanigan and Lewis 2015), and the hilly landscape and beautiful vistas encourage both bicycle and motorcycle touring.

No matter their patrons' mode of transit, local bars become convenient rest spots for visitors and residents alike. Wisconsin has often been ranked the top state for alcohol consumption per capita (Schneider 2015), as bars and their associated heavy weekend consumption are central to community life in the Driftless's small villages. In contrast to other regions, the rural or village bar often also serves as the community's breakfast and coffee spot inasmuch as the local bar or brewery, day or night, serves as a kind of extension of private living spaces in public: "Wisconsin is a tavern state. That is a simple statement of fact, but behind it lies a complex and interesting brew of politics, economics, culture, and social mores, with many great hoppy stories mixed in between" (Draeger and Speltz 2012, 1).

While it is worth noting that alcoholism and binge drinking are a serious problem in Wisconsin (it has been called the "heaviest-drinking state in [the] country"), Wisconsin also has "the most locally-centric alcohol control system in the nation" (Blado 2015; see also Draeger and

Speltz 2012), further strengthening the local-booze connection. And brewing has a long history in Wisconsin. Whereas many locals continue to favor Wisconsin beers from large brewers in Milwaukee (e.g., Pabst and Miller), there is also a significant craft brewery scene across the state. Although most craft breweries are located in or near urban centers, there is often at least one brewery and/or winery in rural counties (fig. 2).

Local brews are clearly favored by many consumers with interests in alternative food networks, but even beyond that subset of the population there is a wider, long-term pride in and commitment to Wisconsin-based industries. In other words, things made or grown in Wisconsin are hip not only for urbanites but for rural residents as well. Breweries market to these consumer preferences in a variety of ways. For example, the largest and most successful craft brewery in Wisconsin is New Glarus, located south of Madison in a village of the same name. New Glarus has been hugely successful in its marketing, thanks to its slogan "Only in WI." Their castle-like brewery and tasting room attracts visitors from many surrounding states, and many of them come to fill up their trunks with cases of beer. This pilgrimage culture is spurred by New Glarus's purposeful local-only distribution tactic, wherein, despite solicitations from (wider) distributors, New Glarus maintains Wisconsin-only distribution (Ruffin 2014). This local focus becomes a draw for people who live nearby and for people who live farther afield.

In addition to a large number of organic farms and numerous breweries, this region is host to a growing number of vineyards and wineries, mostly along the upper Mississippi. These new vineyards are able to grow wine grapes thanks to new varieties created at the University of Minnesota. Although these wineries frequently supplement their cold-weather grapes with imports of grape juice from other wine-growing regions, including New York, Washington, and California, their sales and marketing rely almost exclusively on visits to rural wineries, since few wineries in Wisconsin to date distribute more widely (via grocery or liquor stores, for example).

The (nonexcludable and nonrivalrous) "place-making effect" of the

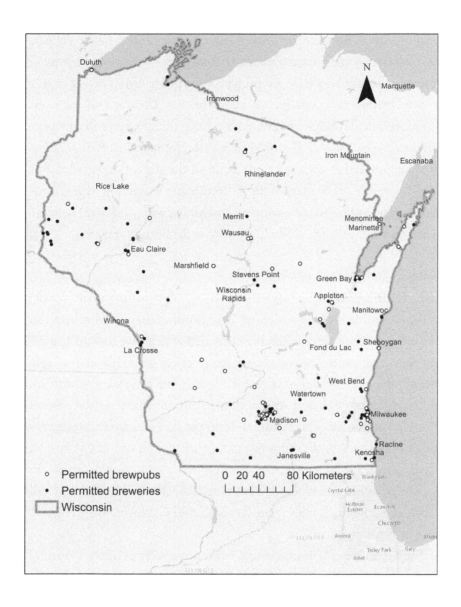

FIG. 2. Locations of brewpubs and breweries in Wisconsin. Cartographer: Shadi Maleki. This map was created using ArcGIS® software by Esri. ArcGIS® and ArcMap™ are the intellectual property of Esri and are used herein under license. Copyright © Esri. All rights reserved. For more information about Esri® software, please visit www .esri.com. The data for this map came from the State of Wisconsin Department of Revenue, dnrmaps.wi.gov, and mapservices.legis.wisconsin.gov.

Driftless area's burgeoning local food movement and booze-related development—with its attendant increases in tourism and business activity—has been experienced primarily as a boon. This is owing largely to the Driftless area's relative isolation: because there is a strong need for economic development yet, so far, little threat of significant gentrification, the food- and booze-related developments in the area have thus far proceeded with relatively little in the way of downsides. For the Midwest or other parts of the United States struggling with serious loss of population, "foodie" tourism (including wine and beer) revenue represents real potential to help otherwise declining rural areas.

Sierra Nevada Wine Country

If the Driftless area can be said to display a "pilgrimage-centric" model of development—wherein sites of local production and consumption of food and booze are created and cultivated with an eye toward attracting tourists keen to enjoy culinary experiences generally unavailable outside the region—our next case study exhibits what we might think of as a different path to a similar destination. In many respects Sierra Nevada wine country is seeking to leverage its existing (and relatively long-established) production infrastructure by refashioning it into an idyllic landscape of consumption suitable for attracting tourists—in particular, tourists disaffected with the arguably more industrialized and urbanized (and inarguably more famous) rival California wine regions, such as Napa and Sonoma.

The Sierra Nevada is home to an extensive wine country spanning eight counties in eastern California. The unique physical characteristics of the region, its rich historical context (site of the famed mid-1800s gold rush), relatively low property values, and excellent growing conditions make the Sierra Nevada an ideal locale to build "wine country" (Myles and Filan 2019). Although the wine region is considered "emerging," it has an extensive history of wine production and is emerging only in relation to the more established California wine regions in Napa and Sonoma Counties. As noted above, area wineries leverage the rural

FIG. 3. Authentic farm life at a vineyard in Amador County, California, in the Sierra Nevada foothills. Photo by Colleen C. Myles, July 2014.

character of this region as a counterpoint to the more commodified and commercialized character of the more well-known wine regions in the state (Myles and Filan 2019). Part of the appeal of the region is the seemingly authentic experience for visitors who wish to engage with the historical context and the production side of the winemaking process (fig. 3). As it is less industrialized and urbanized than other relatively proximate wine regions, it appeals to a segment of the tourism and consumer markets wanting a more personalized experience.

In Sierra Nevada wine country, place and place identity are part of the experience, not just the products being produced. In short, visitors come to consume not just the fermented products of the place but the landscape itself. Rather than attempting to directly compete with Napa and Sonoma, the foothills wineries are choosing an alternative model that is more appropriate to the context of the Sierra Nevada and in turn helps to maintain the character of the region. In this way the Sierra Nevada is making place as it *makes a place* for itself in a changing economic and environmental context (Myles and Filan 2019). This refashioning of the Sierra Nevada's existing winemaking infrastructure is an act of *place-(re)making* that is the joint product of many different wineries' individual business decisions. The impacts of this joint place-(re)making activity redound to all the region's winemakers alike—irrespective of their attitudes (whether favorable, unfavorable, or mixed) to this regional transformation. As with the Driftless area's booze- and food-centered redevelopment, though, the overall effects of this newly created sense of (rural) place have been generally salutary. Because these changes have not resulted in widespread displacement of existing actors (at most they require that existing actors adapt to a region-wide increase in emphasis on the "tourist-facing" aspects of winemaking operations) and because they have not otherwise wrought unduly disruptive effects, the emergence of Sierra Nevada wine country has been regarded in the main as a *good* public good.

The Texas Hill Country Wine Region

In contrast to the Sierra Nevada experience, which profits from the consumption of a popular authenticity, the Hill Country wine region in Texas capitalizes on a (nearly) purely constructed landscape of wine production and consumption. However—and in further contrast to the Sierra Nevada region—the rise of the Hill Country wine industry has forced out many multigeneration farmers and ranchers (UTSOA 2018), meaning the construction and commodification of place has led to rural gentrification in this context. Longtime farm and ranch operations in the Hill Country have been replaced by wineries catering to visitors from surrounding metropolitan areas and beyond. Interestingly, although many wines made in the Hill Country indicate that they are "made in Texas," much of the so-called "Texas wine" marketed there is made with grapes imported from other wine grape growing regions, either elsewhere in Texas (though the supply is relatively limited) or farther afield, such as California, Washington State, Chile, or Argentina (Esco 2016) (fig. 4). A benefit of this importing of grapes is the opportunity

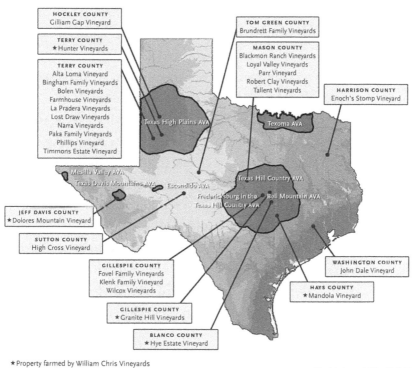

HOCKLEY COUNTY
Gilliam Gap Vineyard

TERRY COUNTY
★ Hunter Vineyards

TERRY COUNTY
Alta Loma Vineyard
Bingham Family Vineyards
Bolen Vineyards
Farmhouse Vineyards
La Pradera Vineyards
Lost Draw Vineyards
Narra Vineyards
Paka Family Vineyards
Phillips Vineyard
Timmons Estate Vineyard

TOM GREEN COUNTY
Brundrett Family Vineyards

MASON COUNTY
Blackmon Ranch Vineyards
Loyal Valley Vineyards
Parr Vineyard
Robert Clay Vineyards
Tallent Vineyards

HARRISON COUNTY
Enoch's Stomp Vineyard

Texas High Plains AVA

Texoma AVA

Mesilla Valley AVA

Texas Davis Mountains AVA   Escondido AVA

Texas Hill Country AVA

Fredericksburg in the   Bell Mountain AVA
Texas Hill Country AVA

JEFF DAVIS COUNTY
★ Dolores Mountain Vineyard

SUTTON COUNTY
High Cross Vineyard

GILLESPIE COUNTY
Fovel Family Vineyards
Klenk Family Vineyard
Wilcox Vineyards

WASHINGTON COUNTY
John Dale Vineyard

HAYS COUNTY
★ Mandola Vineyard

GILLESPIE COUNTY
★ Granite Hill Vineyards

BLANCO COUNTY
★ Hye Estate Vineyard

★ Property farmed by William Chris Vineyards

www.williamchriswines.com | #APieceofOurWorld

FIG. 4. This graphic produced by William Chris Vineyards shows the Texas sources of grapes used in its wines. Photo provided by William Chris Vineyards, Hye, Texas.

for farmers in the Texas Panhandle who had previously relied on water-intensive crops like cotton to diversify into grape growing, particularly the more drought-tolerant species. This benefits the economy and the environment of the Panhandle as farmers become less reliant on mined groundwater from the Ogallala Aquifer (Brown 2012).

Despite the lack of local grape production, every winery in the Hill Country has at least a small vineyard, usually fronting the highway, to promote the idyll of wine country to passersby and visitors. These vineyards, while not contributing much to the wine being produced in the region, do provide the desired image of production for visitors to feel they are experiencing the full spectrum of wine production when they visit the Hill Country. Unlike visitors to California's Sierra Nevada, tourists in the Texas Hill Country are interested in visiting not for what *was* there but for what has been created there now, a (simulated) "wine country" experience produced by the wineries in the region. There is a "sense of place," but it is not necessarily an authentic place or one produced by or rooted in the region's original inhabitants.

Much like urban brewers, Texas Hill Country winemakers, despite the wide-ranging origins of the grapes used, are seen as creating a local product. The act of processing the grapes or grape juice into wine, as well as the subsequent aging of the wine on site, fosters the impression of a locally produced good. In other words, producers may be benefiting, willingly or not, from consumers' assumptions about what makes something local. Producers in the region leverage the elements of production on site, along with a (largely performative) landscape of production, to create for visitors a sense of place that contributes to a growing tourism economy supporting a variety of jobs in the region (Blok 2011; Green 2016).

The Case of Kentucky: Bourbon vs. Brewgrass Trail

The Kentucky Bourbon Trail and the associated Craft Bourbon Tour, which connects primarily rural craft distilleries in Kentucky's Bluegrass region, had nearly nine hundred thousand visitors in 2015, with an average expenditure of $1,000 per visitor (Kentucky Distillers' Association 2016). This makes bourbon tourism a significant contributor to the economic vitality of Kentucky. Tourism is also spurring a building boom as distilleries and communities develop tourism centers and visitor facilities and expand production to cater to growing interest, not just in bourbon consumption, but in witnessing bourbon production (Kentucky Distillers' Association 2016). The production—and consumption in situ—of the fermented product becomes an amenity, drawing visitors and new residents and instigating comprehensive place change, which is often (though not always) perceived as a social and economic benefit. Whether that place change is seen as good or bad, this new place-based identity nevertheless bears the features of nonexcludability and nonrivalry characteristic of public goods.

Although many believe bourbon has to be produced in Kentucky to be called bourbon, it does not; it does, however, have to be produced in the United States, be at least 51 percent corn, and be aged in new charred oak barrels (Mitenbuler 2015). With changing laws and growing demand, bourbon production has expanded to nearly every state (see chapter 4, this volume), although its production remains most closely connected to Kentucky, where more than 90 percent of bourbon production occurs. It is the (strong) connection of the product to a specific place that drives tourism to the Kentucky Bourbon Trail. Visitors want to experience the heritage and history of the product in its birthplace.

Bourbon distillers take advantage of tourists' desire to consume the landscapes of bourbon production by explicitly connecting to place and heritage through marketing (Mitenbuler 2015; Minnick 2016). Distillers incorporate local place-names (e.g., Town Branch Distillery, named for a stream in Lexington), historic individuals (e.g., Elijah Craig, anecdotally credited with inventing bourbon), or even dates (e.g., Barton 1792 Distillery, named for the year Kentucky joined the Union) to establish a Kentucky connection and reinforce the sense of place associated with bourbon (fig. 5). Some of these distillers have legitimate connections to this history, while others simply purchase the rights to historic brands and names as a means to create that connection (see chapter 4). By resurrecting a brand, a new distillery can create an immediate connection to bourbon history and heritage, one that provides an authenticity not available to other new distillers. Consumers respond to this connection, making it a reasonable business decision for new producers to invest in acquiring the rights to these historic brands.

The Brewgrass Trail is a marketing effort of largely Lexington-based craft brewers to capitalize on the popularity of the more well-known rural Bourbon Trail. In contrast to the preceding case studies, the

FIG. 5. Scene from tasting facility at Barton 1792 Distillery, Bardstown, Kentucky. Photo by Christopher Holtkamp, August 2016.

Brewgrass Trail represents an instance of *urban* booze-driven landscape change. As such, it nicely illustrates the potential—latent in all instances of (re)development but less pronounced in the previously examined cases of rural development (save perhaps for the case of the Texas Hill Country)—for place-based public goods to effect various "bads," among them gentrification and race- and class-based stratification.

The Brewgrass Trail includes eight breweries and a collection of "Hop Spots," which are bars and restaurants where patrons can find Kentucky-made beer. Much like the Bourbon Trail, the Brewgrass Trail provides a commemorative passport book that visitors can get stamped at participating venues, and visitors who go to all of the destinations can mail in the completed passport book and receive a free T-shirt. Whereas the Kentucky Bourbon Trail was an effort by select distilleries to help make it easier for bourbon tourists to navigate wide expanses of the Kentucky countryside and visit multiple distilleries, the Brewgrass Trail is much more compact. Six of the breweries are within walking distance of each other, primarily on the west side of downtown Lexington. Rather than celebrating the rural heritage of Kentucky, the Brewgrass Trail is explicitly connecting to spaces and places created through urban revitalization.

The Brewgrass Trail represents competition from the urban craft brewers for the food- and drink-related tourism dollars that have traditionally gone to the rural bourbon distillers. The original marketing video for the trail depicts famous Kentuckian Daniel Boone trekking through the city of Lexington and surrounding areas wearing a raccoon-skin hat, carrying a rifle, and "pounding beer" (a line by the character in the ad), while getting his Brewgrass Trail passport stamped at each of the stops on the trail. The Brewgrass Trail's slogan is "Respect the Bourbon, drink the Beer." By virtue of being in competition with the Bourbon Trail, the Brewgrass Trail makes visible the rural-urban divide, positioning bourbon as a rural product and part of the legacy of Kentucky horse country while situating craft beer as an urban, local product, made and consumed in the city.

Producers recognize the value of coordinating their efforts to create a sense of place related to the concentration of craft brewers in a walkable, urban space (Myles and Breen 2018). In that sort of space, cooperation is intended to produce the proverbial rising tide that lifts all boats, stimulating a virtuous cycle of place creation as the destination becomes desirable for producers and consumers. The transformation of the urban landscape can create positive externalities by generating increased tax revenues and physical improvements to what may have been a declining neighborhood. Although the businesses and neighborhood are open and accessible to all, it may be that not all feel welcome, since the businesses largely cater to a middle- or upper-middle-class clientele (and in this case serve beverages that can be legally consumed only by persons at least twenty-one years of age). Additionally, lower-income residents may be forced to relocate when property values and rents increase as the area transforms and, by many accounts, the neighborhood gentrifies (Mathews and Picton 2014). It is critical to recognize the potential negative externalities of (craft) fermentation-focused development, even as these industries are hailed for their capacity to (positively) change communities. Again, whether perceived as beneficial or not, such community changes can be characterized as a public good in that their impacts are felt by all (i.e., are nonexcludable) and, because one person's experience does not diminish someone else's, are also nonrival.

Discussion

The Brewers Association (n.d.) has reported an economic impact of craft brewing that totaled $67.8 billion, and that activity alone was responsible for more than 456,000 jobs. As of August 2017 nearly 20,000 people were directly employed in 1,589 craft distilleries operating in the United States (American Craft Spirits Association 2017).[3] In Kentucky alone as of 2014 agricultural production directly serving distilleries accounted for 1,360 jobs and $56 million in value, along with another 100 jobs and $2 million in agricultural support jobs (Kornstein and Luckett 2014). As craft production continues to gain popularity, the economic impacts will continue to increase as well. Increased demand for small-scale local production may contribute to more farmers adopting methods to serve that demand. Moreover, because consumers are interested in knowing the life cycle of the products they are purchasing, the relationship between those producing the raw materials and the final product will become more important (Morgan, Marsden, and Murdoch 2009). In addition to its direct economic impacts, craft production is also a significant driver of tourism (Kline, Slocum, Cavaliere 2017; Slocum, Kline, Cavaliere 2018).

As presented above, craft producers are active agents in the creation of place in both rural and urban environments, where they foster both positive and negative externalities by their choice of locations and the changes in the physical and cultural landscapes produced by those choices. Producers recognize the value of connection to place and both benefit from and contribute to changes in the places where they choose to locate. These changes include not only private investment by businesses locating in a place but also public investment by governmental agencies responding to changes in an area and seeking to capitalize on those changes (Manzo and Perkins 2006). The outcome of private and public investments results in the creation of place, a geographic area where people actively choose to spend time (and money) because of the physical, social, and cultural attributes of that place. This place attachment contributes to the success of businesses associated with the area.

Such place-making, which contributes to economic growth and neighborhood change, is a public good (nonexcludable, nonrival), but the good is not necessarily "good" for everyone. In other words, the benefits accrued from the creation of place are not evenly distributed. As the places created by craft producers often cater to an upper-middle-class, mostly white clientele, some people are left out of the good being created by the rise of craft fermentation. Thus, while not actively excluding anyone, the spaces created may not make people of color or lower-income persons feel entirely welcome.

In rural areas craft producers can transform the landscape so as to attract visitors from near or far. The Texas Hill Country has been transformed from terrain covered with so-called cedars (actually ashe juniper), and primarily suited for raising goats, into wine country, which has a major tourism component. Despite the fact that the vast majority of grapes used to produce Hill Country wine are actually grown in the Texas Panhandle (Texas Wine and Grape Growers Association n.d.), most wineries in Central Texas maintain on-site vineyards to help create the pastoral image desired by visitors. Whereas Texas wineries are making a largely symbolic connection between grape growing and wine production, producers in the Sierra Nevada have an authentic relationship between the two. Either way, the landscape becomes as much a commodity to be consumed as the wine being produced (fig. 6). Visitors want to believe that they are seeing the whole production cycle from grape to glass, and producers are feeding that desire through the creation of place, even if it is a simulacrum of reality, as in the case of the Texas Hill Country wineries.

In urban spaces craft producers can more easily be viewed as creators

FIG. 6. Picture perfect: a frame-within-a frame image of the William Chris Vineyards in the Texas Hill Country, where vineyard scenery is as much a commodity as the wines produced there. Photo by Colleen C. Myles, November 2015.

of place—and as agents in gentrification (Mathews and Picon 2014; Reid and Gatrell 2017). As discussed above, craft producers often choose to locate in previously less desirable (and thus less expensive locations), and by locating in these places they immediately begin to change them (Patterson and Hoalst-Pullen 2014; Gatrell, Reid, and Steiger 2018; Slocum, Kline, and Cavaliere 2018). Other producers typically follow the early adopters, and complementary businesses soon follow (Nilsson, Reed, and Lehnert 2017). Previously neglected or ignored places become appealing destinations because of the producers located in these spaces (Myles and Breen 2018). Additionally, the producers see value in connecting to that place because of consumers' desire to connect through their consumption. The result is often rising land and building prices, increasing rents, and, ultimately, shifting neighborhood composition and form. While such transitions can force out longtime residents, for some the neighborhood change is seen as normatively good, since residents may be able to capitalize on the increased property values. Many, however—especially renters and lower-income residents and lower-value businesses—may lose access to that place and might be forced to relocate. In sum, while craft producers are certainly creating places, those places are not always welcoming for everyone. It is precisely because of its uneven benefits that we frame fermentation-focused place-making as a public good—because the changes affect all, whether they like it or not.

Conclusion

Craft fermenters are agents in the creation of place, which can be considered a public good, with the caveats discussed above. Whether it is a rural environment with lovely views of bucolic wine country or a thriving urban neighborhood with trendy breweries and shops, these are places where people want to be and spend time. The fermented landscapes created by these producers become part of the experience for visitors, who come to these places to interact with the landscape and potentially with other people sharing that space with them. The created place is open and accessible to all, even if it's not the case that everyone always feels welcome, as we have discussed.

Additionally, craft fermenters contribute to the public good through the creation of jobs, increased tax revenues, and physical improvements in neighborhoods as a result of the opening of these new enterprises. Job creation includes not only direct employment at the individual businesses but also spin-off jobs created by suppliers providing raw materials, as well as other actors who are part of the craft business ecosystem; there is thus a multiplier effect for localized craft production. These positive externalities accrue to the larger community, even if the craft product itself is not consumed by all.

The clear detriment (negative externalities) of the public good created by the craft fermentation industry is the noninclusive nature of the places being created. As discussed, craft producers are often the harbingers of gentrification, serving as a sign that rising rents and property values are soon to follow. Longtime residents and businesses with smaller profit margins often fall victim to rising costs as neighborhoods transform around them, sometimes turning into places where they no longer are welcome. While these detriments to certain subpopulations might suggest that the term "public good" is inappropriate (namely because its effect is not "good"), the "good" in the term "public good" is not normative; it is an economic unit, a descriptor.[4]

When a craft producer locates in a neighborhood, followed by other producers and other complementary businesses, that neighborhood will inevitably change. The place created through resulting urban or rural (re) development is "public" and thus is a public good, even if the outcomes (e.g., rising rents and property values) are not universally positive. The owners of these new businesses are striving to gain economic benefits for themselves individually, but they also recognize the additional gains to be had through the creation of place, wherein property values increase and businesses have greater viability and profitability (Myles and Breen 2018).

As discussed, craft fermentation industries often occupy spaces created by, and for, upper-middle-class consumers. And, despite growing inclusivity (Watson 2014), the industry remains largely the realm of white people. Therefore, the places being created by craft producers are not always inclusive spaces, but the attendant economic changes are not exclusive either. In fact many of the challenges noted here result from the universality of the economic shifts. Specifically, although the economic benefits of craft production are significant, the relationship to place-making is more problematic. By creating demand for place and products connected to place, craft producers are providing a benefit to some businesses through new opportunities to produce goods and services for craft producers. They are also contributing to the creation of successful places by serving as the first wave, investing in low-cost, less desirable neighborhoods. This investment contributes to those places becoming more desirable over time. In this way, although the craft fermentation businesses and places in which those businesses are located may not be explicitly exclusionary, the products being produced and the overall sense of place may not be welcoming or beneficial to all members of society; nevertheless, craft booze producers are creating a public good in the technical sense of the term. Establishing the nonexcludable and nonrival nature of such public goods is thus an essential start to figuring out how to ensure that the resulting spaces and places created via fermentation-focused industry are more reliably *good*.

Notes

This chapter is the result of a lively conversation and collaboration between scholars at several different institutions, and it could not have been written without thoughtful preparatory discussion with Jessi Breen. Furthermore, the assistance of Shadi Maleki in the final manuscript preparations was indispensable. Thanks also to the organizers and attendees of the Philosophy, Politics, and Economics (PPE) Society meeting (New Orleans, March 2018) for providing an interdisciplinary forum for this work's first public hearing.

1. Increasing interest in locally produced goods has led consumers to actively become producers themselves, as demonstrated by the distinct rise in "craft culture," home-brewing, and other home production hobbies, as well as a resurgence of "old jobs," including artisanal commercial crafters (Katz 2006; Levine and Heimerl 2008; Ocejo 2017).

2. For the purpose of this research, we adopt this definition of "craft": a "craft producer is someone who exercises personal control over all the processes involved in the manufacture of the good in question" (Campbell 2005, 27).

3. Although increased production and associated tourism provide significant revenue and employment, many of these jobs are low-wage and seasonal (Wang and Pfister 2008).

4. Readers still skeptical as to the propriety of the term "public good" in this context—as a descriptor for items not universally regarded as good or valuable—should note that their skepticism would likewise militate against all instances of the term "good" in economics, for few (if any) paradigmatically private goods are thus universally regarded. Consider the case of cigarettes, or quinoa, or Cheetos, or kale, or, for that matter, pints of beer and bottles of wine. Surely there are parties who do not regard these items as good, but this fact doesn't undermine our understanding that they are among the "goods and services" produced in our economy.

References

Alkon, Alison Hope. 2012. *Black, White, and Green: Farmers Markets, Race, and the Green Economy*. Athens: University of Georgia Press.

Alkon, Alison Hope, and Julian Agyeman, eds. 2011. *Cultivating Food Justice: Race, Class, and Sustainability*. Cambridge MA: MIT Press.

American Craft Spirits Association. 2017. "Craft Spirits Producers Sold Nearly 6 Million Cases Last Year Alone." October 24, 2017. https://americancraftspirits.org/craft-spirits-producers-sold-nearly-6-million-cases-last-year-alone/.

Arbel, Tali. 2013. "Build a Craft Brewery, Urban Revival Will Come." *USA Today*, July 6, 2013. https://www.usatoday.com/story/money/business/2013/07/05/in-urban-revival-beer-creates-small-business-hubs/2487625/.

Argent, Neil, Matthew Tonts, Roy Jones, and John Holmes. 2013. "A Creativity-Led Rural Renaissance? Amenity-Led Migration, the Creative Turn and the Uneven Development of Rural Australia." *Applied Geography* 44 (October): 88–98. https://doi.org/10.1016/j.apgeog.2013.07.018.

Atkins, Peter, and Ian Bowler. 2001. *Food in Society: Economy, Culture, Geography*. London: Arnold.

Barajas, Jesus M., Geoff Boeing, and Julie Wartell. 2017. "Neighborhood Change, One Pint at a Time: The Impact of Local Characteristics on Craft Breweries." *SSRN Electronic Journal*. Last revised July 29, 2017. https://doi.org/10.2139/ssrn.2936427.

Beebe, Craig, and Stephen M. Wheeler. 2012. "Gold Country: The Politics of Landscape in Exurban El Dorado County, California." *Journal of Political Ecology* 19 (1). https://doi.org/10.2458/v19i1.21710.

Beggs, Jodi. 2018. "Private Goods, Public Goods, Congestible Goods, and Club Goods." *ThoughtCo*. Accessed March 29, 2018. https://www.thoughtco.com/excludability-and-rivalry-in-consumption-1147876.

Best, Allen. 2015. "Welcome to Beer Country." American Planning Association, February 2015. https://www.townofmaynard-ma.gov/wp-content/uploads/2015/03/edc-welcome-to-beer-country-APA.pdf.

Blado, Kayla. 2015. "Wisconsin Heaviest-Drinking State in Country, Study Finds." *Wisconsin Public Radio*, April 29, 2015. https://www.wpr.org/wisconsin-heaviest-drinking-state-country-study-finds.

Blok, Celestina. 2011. "Deep in the Heart of Texas Wines." *Fort Worth Business Press* 23 (10): 31.

Brewers Association. n.d. "The Craft Brewing Industry Contributed $67.8 Billion to the U.S. Economy in 2016, More Than 456,000 Jobs." Accessed March 27, 2018. https://www.brewersassociation.org/statistics/economic-impact-data/.

Brown, Karen. 2012. "Amid Drought, Wine Grapes Save a Cotton Farmer." *CBS News*, July 31, 2012. https://www.cbsnews.com/news/amid-drought-wine-grapes-save-a-cotton-farmer/.

Campbell, Colin. 2005. "The Craft Consumer: Culture, Craft and Consumption in a Postmodern Society." *Journal of Consumer Culture* 5 (1): 23–42. https://doi.org/10.1177/1469540505049843.

Chapman, Nathaniel G., J. Slade Lellock, and Cameron D. Lippard. 2017. *Untapped: Exploring the Cultural Dimensions of Craft Beer*. Morgantown: West Virginia University Press.

Chase, Jacquelyn. 2015. "Bending the Rules in the Foothills—County General Planning in Exurban Northern California." *Society & Natural Resources* 28 (8): 857–72. https://doi.org/10.1080/08941920.2015.1045643.

City of Portland. n.d. "Craft Fermented Beverage Industry." Environmental Services. Accessed January 23, 2018. https://www.portlandoregon.gov/bes/article/650287.

Cornes, Richard, and Todd Sandler. 1996. *The Theory of Externalities, Public Goods and Club Goods*. Cambridge: Cambridge University Press.

Crawford, Margaret. 1992. "The World in a Shopping Mall." In *Variations on a Theme Park: The New American City and the End of Public Space*, edited by Michael Sorkin, 3–30. New York: Hill and Wang.

Davis, Genie. 2015. "What Makes a Beverage 'Craft?'" *CrushBrew Magazine*, September 10, 2015. http://crushbrew.com/about-crushbrew.

Donald, Betsy. 2009. "From Kraft to Craft: Innovation and Creativity in Ontario's Food Economy." Working paper 2009-WPONT-001. Martin Prosperity Institute, Rotman School of Management, University of Toronto. http://martinprosperity.org/media/pdfs/From_Kraft_to_Craft-B_Donald.pdf.

Dawson, John, and Steve Burt. 1998. *European Retailing: Dynamics, Restructuring, and Development Issues*. London: Routledge.

Draeger, Jim, and Mark Speltz. 2012. *Bottoms Up: A Toast to Wisconsin's Historic Bars and Breweries*. Madison: Wisconsin Historical Society Press.

Driftless Wisconsin. n.d. "The Heart of Wisconsin's Driftless Area." *Driftless Wisconsin Region Guide*. Accessed April 5, 2018. https://driftlesswisconsin.com/.

Duane, Timothy P. 1999. *Shaping the Sierra: Nature, Culture, and Conflict in the Changing West*. Berkeley: University of California Press.

Esco, Melinda. 2016. "A Look at the Big Business of Texas Wine, including a History of Grape Growing and Winemaking." In *Texas Almanac 2016–2017*, edited by Elizabeth Cruze Alvarez and Robert Plocheck, 38–45. Austin: Texas State Historical Association.

Flack, Wes. 1997. "American Microbreweries and Neolocalism: 'Ale-Ing' for a Sense of Place." *Journal of Cultural Geography* 16 (2): 37–53. https://doi.org/10.1080/08873639709478336.

Flanigan, Kathy, and Chelsea Lewis. 2015. "A Beer Tour of Wisconsin's Driftless Region: Touring and Tasting in Southwestern Wisconsin." *Journal Sentinel* (Milwaukee), July 9, 2015. http://archive.jsonline.com/entertainment/beer/a-beer-tour-of-wisconsins-driftless-region-b99531104z1-312635931.html.

Garavaglia, Christian, and Johan Swinnen. 2018. "Economics of the Craft Beer Revolution: A Comparative International Perspective." In *Economic Perspectives on Craft Beer: A Revolution in the Global Beer Industry*, edited by Christian Garavaglia and Johan Swinnen, 3–51. Cham, Switzerland: Palgrave Macmillan.

Gatrell, Jay, Neil Reid, and Thomas L. Steiger. 2018. "Branding Spaces: Place, Region, Sustainability and the American Craft Beer Industry." *Applied Geography* 90 (January): 360–70. https://doi.org/10.1016/j.apgeog.2017.02.012.

Gosnell, Hannah, and Jesse Abrams. 2009. "Amenity Migration: Diverse Conceptualizations of Drivers, Socioeconomic Dimensions, and Emerging Challenges." *GeoJournal* 76 (4): 303–22. https://doi.org/10.1007/s10708-009-9295-4.

Green, Justin. 2016. "Popping the Cork on Texas' Wine Industry." *Texas Agriculture Magazine* 31 (13): 16.

Hayward, Scott D., and David Battle. 2017. "Brewing a Beer Industry in Asheville, North Carolina." *Craft Beverages and Tourism* 2 (12): 171–93. https://doi.org/10.1007/978-3-319-57189-8_11.

Hindy, Steve, and John Hickenlooper. 2015. *The Craft Beer Revolution: How a Band of Microbrewers Is Transforming the World's Favorite Drink*. New York: Palgrave Macmillan.

Hiner, Colleen C. 2014. "'Been-Heres vs. Come-Heres' and Other Identities and Ideologies along the Rural-Urban Interface: A Comparative Case Study in Calaveras County, California." *Land Use Policy* 41 (1): 70–83. https://doi.org/10.1016/j.landusepol.2014.05.001.

Holtkamp, Chris, Thomas Shelton, Graham Daly, Colleen C. Hiner, and Ronald R. Hagelman. 2016. "Assessing Neolocalism in Microbreweries." *Papers in Applied Geography* 2 (1): 66–78. https://doi.org/10.1080/23754931.2015.1114514.

Hopkins, David. 2014. "How Craft Beer (Finally) Came to Dallas." *D Magazine*, June 2014. https://www.dmagazine.com/publications/d-magazine/2014/june/dallas-first-microbreweries/.

Hubbard, Phil. 2016. "Hipsters on Our High Streets: Consuming the Gentrification Frontier." *Sociological Research Online* 21 (3): 1–6. https://doi.org/10.5153/sro.3962.

Ilbery, Brian, Carol Morris, Henry Buller, Damian Maye, and Moya Kneafsey. 2005. "Product, Process and Place." *European Urban and Regional Studies* 12 (2): 116–32. https://doi.org/10.1177/0969776405048499.

Infante, Dave. 2015. "There Are Almost No Black People Brewing Craft Beer. Here's Why." *Thrillist*, December 3, 2015. https://www.thrillist.com/drink/nation/there -are-almost-no-black-people-brewing-craft-beer-heres-why.

Jordan, Jennifer. 2016. "Drinking Revolution, Drinking in Place: Craft Beer, Hard Cider, and the Making of North American Landscapes." Presentation at the Dublin Gastronomy Symposium, Dublin Institute of Technology, Dublin, May 31–June 1, 2016. https://arrow.dit.ie/cgi/viewcontent.cgi?article=1088&context=dgs.

Katz, Sandor Ellix. 2006. *The Revolution Will Not Be Microwaved: Inside America's Underground Food Movements*. White River Junction VT: Chelsea Green.

Kentucky Distillers' Association. 2016. "Kentucky Bourbon Trail® Visits Skyrocket with 900,000 Guests in 2015." Kentucky Distillers' Association Media Center, January 21, 2016. https://kybourbon.com/kentucky-bourbon-trail-visits-skyrocket -with-900000-guests-in-2015/.

Kline, Carol, Susan L. Slocum, and Christina T. Cavaliere. 2017. *Craft Beverages and Tourism, Volume 1: The Rise of Breweries and Distilleries in the United States*. Cham, Switzerland: Springer International.

Kornstein, Barry, and Jay Luckett. 2014. "The Economic and Fiscal Impacts of the Distilling Industry in Kentucky." Kentucky Agricultural Development Fund, October 2014. http://kybourbon.com/wp-content/uploads/2014/08/economic_impact _2014.pdf.

Larsen, Soren, and Craig Hutton. 2011. "Community Discourse and the Emerging Amenity Landscapes of the Rural American West." *GeoJournal* 77 (5): 651–65. https://doi.org/10.1007/s10708-011-9410-1.

Lees, Loretta, Tom Slater, and Elvin Wyly. 2008. *Gentrification*. New York: Routledge.

Levine, Faythe, and Cortney Heimerl. 2008. *Handmade Nation: The Rise of DIY, Art, Craft, and Design*. New York: Princeton Architectural Press.

Manzo, Lynne C., and Douglas D. Perkins. 2006. "Finding Common Ground: The Importance of Place Attachment to Community Participation and Planning." *Journal of Planning Literature* 20 (4): 335–50. https://doi.org/10.1177 /0885412205286160.

Mathews, Vanessa, and Roger M. Picton. 2014. "Intoxifying Gentrification: Brew Pubs and the Geography of Post-Industrial Heritage." *Urban Geography* 35 (3): 337–56. https://doi.org/10.1080/02723638.2014.887298.

Maye, Damian, Lewis Holloway, and Moya Kneafsey. 2007. *Alternative Food Geographies: Representation and Practice*. Amsterdam: Emerald.

McKeithan, Seán S. 2012. "Every Ounce a Man's Whiskey? Bourbon in the White Masculine South." *Southern Cultures* 18 (1).

McKinnon, Innisfree. 2016. "Mixing Productive and Consumptive Economies: Farmer Perspectives on Agriculture and Tourism in Southwestern Wisconsin." Unpublished paper.

Minnick, Fred. 2016. *Bourbon: The Rise, Fall, and Rebirth of an American Whiskey*. Minneapolis: Voyageur Press.

Mitenbuler, Reid. 2015. *Bourbon Empire: The Past and Future of America's Whiskey*. New York: Penguin Books.

Morgan, Kevin, Terry Marsden, and Jonathan Murdoch. 2009. *Worlds of Food: Place, Power, and Provenance in the Food Chain*. Oxford: Oxford University Press.

Myles, Colleen C., and Jessica McCallum Breen. 2018. "(Micro) Movements and Microbrew: On Craft Beer, Tourism Trails, and Material Transformations in Three Urban Industrial Sites." In *Economic Perspectives on Craft Beer: A Revolution in the Global Beer Industry*, edited by Christian Garavaglia and Johan Swinnen, 159–70. Cham, Switzerland: Palgrave Macmillan.

Myles, Colleen C., and Trina R. Filan. 2017. "Boom-and-Bust: (Hi)stories of Landscape Production and Consumption in California's Sierra Nevada Foothills." *Polymath: An Interdisciplinary Arts and Sciences Journal* 7 (2): 76–89.

Myles, Colleen C., and Trina Filan. 2019. "Making (a) Place: Wine and the Production and Consumption of Landscape in the Sierra Nevada Foothills." *Regional Studies, Regional Science* 6 (1): 157–67.

Nahapiet, Janine, and Sumantra Ghoshal. 1998. "Social Capital, Intellectual Capital, and the Organizational Advantage." *Academy of Management Review* 23 (2): 242–66.

Nilsson, Isabelle, Neil Reid, and Matthew Lehnert. 2017. "Geographic Patterns of Craft Breweries at the Intraurban Scale." *Professional Geographer* 70 (1): 114–25. https://doi.org/10.1080/00330124.2017.1338590.

Nixon, Rob. 2000. "Navigating through the Driftless." *New York Times*, September 10, 2000.

Ocejo, Richard E. 2017. *Masters of Craft: Old Jobs in the New Urban Economy*. Princeton: Princeton University Press.

Orrick, Dave. 2016. "Viroqua: Mecca for Driftless Stream Trout." *Pioneer Press* (St. Paul MN), TwinCities.com. Last updated April 26, 2016. https://www.twincities.com/2016/04/25/viroqua-wi-driftless-stream-trout-fly-fishing-mecca/.

Patterson, Mark, and Nancy Hoalst-Pullen, eds. 2014. *The Geography of Beer: Regions, Environment, and Societies*. New York: Springer.

Perritt, Marcia Machado. 2013. "Breweries and Economic Development: A Case of Home Brew." *Community and Economic Development Blog*, UNC School of Gov-

ernment, April 5, 2013. https://www.sog.unc.edu/blogs/community-and-economic
-development-ced/breweries-and-economic-development-case-home-brew.

Putnam, Robert. 1993. "The Prosperous Community: Social Capital and Public Life."
American Prospect 13 (4): 35–42.

Reid, Neil, and Jay D. Gatrell. 2017. "Craft Breweries and Economic Development:
Local Geographies of Beer." *Polymath: An Interdisciplinary Arts and Sciences Journal*
7 (2): 90–110.

Ruffin, Josh. 2014. "New Glarus Brewery Is Hell Bent on Not Taking Over the World."
Paste Magazine, November 6, 2014. https://www.pastemagazine.com/articles/2014
/11/new-glarus-brewery-is-hell-bent-on-not-taking-over.html.

Schneider, Doug. 2015. "Binge Drinking in State Still Far Exceeds U.S. Average." *Green
Bay (WI) Press Gazette*, April 23, 2015. https://www.greenbaypressgazette.com
/story/news/local/2015/04/23/binge-drinking-wis-still-far-exceeds-us-average
/26268831/.

Schnell, Steven M., and Joseph F. Reese. 2003. "Microbreweries as Tools of Local
Identity." *Journal of Cultural Geography* 21 (1): 45–69. https://doi.org/10.1080
/08873630309478266.

———. 2014. "Microbreweries, Place, and Identity in the United States." In *The Geography of Beer: Regions, Environment, and Societies*, edited by Mark Patterson and
Nancy Hoalst-Pullen, 167–87. New York: Springer.

Senese, Donna, Filippo Randelli, John S. Hull, and Colleen C. Myles. 2018. "Drinking
in the Good Life: Tourism Mobilities and the Slow Movement in Wine Country."
In *Slow Tourism, Food and Cities: Pace and the Search for the "Good Life,"* edited by
Michael Clancy, 214–31. New York: Routledge.

Shears, Andrew. 2014. "Local to National and Back Again: Beer, Wisconsin and Scale."
In *The Geography of Beer: Regions, Environment, and Societies*, edited by Mark
Patterson and Nancy Hoalst-Pullen, 154–63. New York: Springer.

Shoup, Mary Ellen. 2017. "US Craft Spirit Industry on the Rise as Producers Double on-
Premise Investments." *Beverage Daily*, October 27, 2017. https://www.beveragedaily
.com/Article/2017/10/27/US-craft-spirit-industry-on-the-rise-as-producers-double
-on-premise-investments.

Simonson, Robert. 2018. "Craft Distillers, Facing Lower Taxes, Invest in Themselves."
New York Times, April 24, 2018, D3.

Sims, Rebecca. 2009. "Food, Place and Authenticity: Local Food and the Sustainable
Tourism Experience." *Journal of Sustainable Tourism* 17 (3): 321–36. https://doi.org
/10.1080/09669580802359293.

SIRC. 2000. *Social and Cultural Aspects of Drinking.* Social Issues Research Centre.
http://www.sirc.org/publik/drinking6.html.

Slocum, Susan L., Carol Kline, and Christina T. Cavaliere. 2018. *Craft Beverages and Tourism, Volume 2: Environmental, Societal, and Marketing Implications*. Cham, Switzerland: Springer International.

Spracklen, Karl, Jon Laurencic, and Alex Kenyon. 2013. "Mine's a Pint of Bitter: Performativity, Gender, Class and Representations of Authenticity in Real-Ale Tourism." *Tourist Studies* 13 (3): 304–21. https://doi.org/10.1177/1468797613498165.

Stopa, Alex. 2016. "How Niche Products Are Transforming the Beverage Industry." *My Process Expo*, January 14, 2016. http://www.myprocessexpo.com/blog/beverage/how-niche-products-are-transforming-the-beverage-industry/.

Taylor, Kate. 2016. "The Battle between Big Beer and Craft Brewers Is Getting Ugly." *Business Insider*, February 11, 2016. http://www.businessinsider.com/big-beer-vs-craft-beer-battle-gets-ugly-2016-2.

Taylor, Laura. 2011. "No Boundaries: Exurbia and the Study of Contemporary Urban Dispersion." *GeoJournal* 76 (4): 323–39. https://doi.org/10.1007/s10708-009-9300-y.

Taylor, Laura Elizabeth, and Patrick T. Hurley, eds. 2016. *A Comparative Political Ecology of Exurbia: Planning, Environmental Management, and Landscape Change*. Cham, Switzerland: Springer International.

Texas Wine and Grape Growers Association. n.d. "Texas Wine Industry Facts." Accessed March 22, 2018. https://www.txwines.org/about-texas-wine

Thompson, Derek. 2018. "Craft Beer Is the Strangest, Happiest Economic Story in America." *The Atlantic*, January 19, 2018. https://www.theatlantic.com/business/archive/2018/01/craft-beer-industry/550850/?utm_source=atlfb.

Thurnell-Read, Thomas. 2016. "The Embourgeoisement of Beer: Changing Practices of 'Real Ale' Consumption." *Journal of Consumer Culture* 18 (4): 539–57. https://doi.org/10.1177/1469540516684189.

Travis, William Riebsame. 2007. *New Geographies of the American West Land Use and the Changing Patterns of Place*. Washington DC: Island Press.

Tuan, Yi-Fu. 1991. "Language and the Making of Place: A Narrative-Descriptive Approach." *Annals of the Association of American Geographers* 81 (4): 684–96. https://doi.org/10.1111/j.1467-8306.1991.tb01715.x.

UTSOA (University of Texas School of Architecture). 2018. "Toward a Regional Plan for the Texas Hill Country." Accessed March 3, 2018. https://soa.utexas.edu/sites/default/disk/Toward-a-Regional-Plan-for-the-Texas-Hill-Country.pdf.

Wagner, Brock, and Scott Metzger. 2017. "Big Beer Trying to Swallow Craft Beer Industry." *San Antonio Express-News*, December 18, 2017. https://m.mysanantonio.com/opinion/commentary/article/Big-Beer-trying-to-swallow-craft-beer-industry-12434665.php.

Walker, Peter, and Louise Fortmann. 2003. "Whose Landscape? A Political Ecology of the Exurban Sierra." *Cultural Geographies* 10 (4): 469–91. https://doi.org/10.1191/1474474003eu285oa.

Wang, Yasong Alex, and Robert E. Pfister. 2008. "Residents' Attitudes toward Tourism and Perceived Personal Benefits in a Rural Community." *Journal of Travel Research* 47 (1): 84–93. https://doi.org/10.1177/0047287507312402.

Watson, Bart. 2014. "The Demographics of Craft Beer Lovers." Presentation at the Great American Beer Festival, Brewers Association, Denver CO, October 3, 2014. https://www.brewersassociation.org/wp-content/uploads/2014/10/Demographics-of-craft-beer.pdf.

Zasada, Ingo. 2011. "Multifunctional Peri-Urban Agriculture—A Review of Societal Demands and the Provision of Goods and Services by Farming." *Land Use Policy* 28 (4): 639–48. https://doi.org/10.1016/j.landusepol.2011.01.008.

Zelinsky, Wilbur. 2011. *Not Yet a Placeless Land: Tracking an Evolving American Geography*. Amherst: University of Massachusetts Press.

Zukin, Sharon. 1991. *Landscapes of Power: From Detroit to Disney World*. Berkeley: University of California Press.

———. 2011. "Reconstructing the Authenticity of Place." *Theory and Society* 40 (2): 161–65. https://doi.org/10.1007/s11186-010-9133-1.

Landscapes of Failure 3

Why Do Some Wine Regions Not Succeed?

John Overton

Landscapes of Success

The expansion and transformation of the global wine industry in the past four decades has transformed landscapes in many parts of the world. With globalization, there has been a move toward more differentiated wine production (alongside a mass market for cheaper bulk wine), with emphasis given to variety, place of origin, putative quality, vintage, and artisanal methods. Existing wine regions have been transformed and expanded, with the growth of winery restaurants, tours, cellar door sales, and the switch to supposed higher-quality grape varieties rather than bulk production. A significant result of such activity is that new wine regions have emerged in many parts of the world, including Europe, the New World, and even the so-called Third World (Banks and Overton 2010). Here again we see new wine landscapes being created and replacing former rural land uses with intensive grape production and associated sites of consumption (restaurants and wine tasting) and retailing (wine and wine paraphernalia). These new wine landscapes seek to attract customers and promote images of sophistication, wealth, quality, and distinction (Demossier 2004, 2005; Ulin 1995, 2007; Charters 2006).

The development of these wine regions has attracted much attention from scholars. Books, articles in the *Journal of Wine Research*, and conference papers have documented and analyzed the way these regions have grown, what they have produced, and how they have transformed local geographies. The approach has been mostly triumphal: wine regions have grown and become oases of success and sophistication in rural regions that it might be assumed are otherwise faltering (see chapter 2, this volume). The new wine landscapes have been distinctive and new, producing an image of neatness and order and presenting a strange combination of bucolic nostalgia (the bounty, fresh air, and craft of the countryside that draws on centuries-old winemaking traditions) mixed with modernity and global connectedness (innovative winemaking techniques, export, and international recognition).[1] Such is the approach that I have employed in studies of wine regions in Chile and New Zealand (Overton, Murray, and Pino Silva 2012; Overton and Heitger 2008; Murray and Overton 2011), and there are many other examples (see, e.g., Banks and Sharpe 2006; Bramble et al. 2007; Sánchez-Hernández, Aparicio-Amador, and Alonso-Santos 2010; Swart and Smit 2009; and Morris 2000). Scholars have documented and celebrated apparent success, and we have seen wine as the salient for rural transformation and landscape enhancement. To link to the theme of this book, we have traced the way wine regions have gone through a process of "fermentation" and produced fine landscapes as well as fine wine (see chapter 1, this volume). However, we have been largely silent on failure. We have not documented the way some wine regions have failed to develop or have gone through a process of contraction and decline.

One of the enduring themes in studies of the way wine regions have developed and succeeded has been the concept of *terroir*. This describes the collection of environmental and cultural factors (soils, climate, geology, topography, aspect, traditions, history, and culture) that define unique places (and regions) but also imbue wines with apparently distinctive characteristics and link the qualitative aspects of wine to place (Wilson 1998; Fanet 2004; Sommers 2008). Terroir seems to provide a framework for understanding why some places—and landscapes—emerge as fine wine regions, though in the literature there seems to be a rather post hoc analytical approach (looking back to explain the success of an established region). In addition, although terroir allows for human elements, it remains based on a strong foundation of environmental determinism: soils and climate define wine quality and thus wine landscapes. Both critics and skeptics of the terroir concept see it as a mechanism that is employed in ways to construct place and develop narratives of quality and distinctiveness that can promote and add value to commodities (Moran 2001; Barham 2003; Overton and Murray 2016). It also has the potential to be employed to promote some interests and land uses over others in cases when there are "competing rural capitalisms" (Taylor and Hurley 2016, 8).

Other analytical frameworks that might be employed to examine the way regions develop include those from economic geography, notably the interest in industrial clusters (Giuliani, Pietrobelli, and Rabelotti 2005; Charters and Michaux 2014; Fløysand, Jakobsen, and Bjarnar 2012), and some rather established notions of economic rent, which acknowledge competing rural land uses. Another point to remember is that some popular literature—as well as part of the terroir approach— also makes reference to human agency, particularly the role of pioneering individuals who experiment, persevere, survive setbacks, take risks, and heroically achieve success against the odds (Thomson 2012). All these approaches will help to address the simple yet thus far neglected question of why some wine regions do not succeed.

To this end, this chapter examines the case of the Canterbury Province in the South Island of New Zealand. Two adjacent wine regions have emerged there. One—Waipara—has grown and become established— and received academic attention (Schuster, Jackson, and Tipples 2002; Tipples 2007; Overton, Banks, and Murray 2014), but the other—the scattered vineyards to be found on the Canterbury Plains and the hills of Banks Peninsula close to the city of Christchurch)—has failed to develop and has been largely neglected.

Materials and Methods

This chapter concentrates on the Canterbury Plains and Banks Peninsula winegrowing areas of the wider Canterbury Region and contrasts these with the Waipara winegrowing district, also within Canterbury (fig. 7).[2]

It draws on the author's long interest in the geography of New Zealand wine (Overton 1996; Overton and Heitger 2008; Overton and Murray 2014b; Overton, Banks, and Murray 2014) and from close observation of the emerging Canterbury wine industry in the 1990s. It also employs data from vineyard surveys collected by the industry body New Zealand Winegrowers, as well as website information from the region's wineries and a number of secondary sources on the region's development. It interrogates the narratives of history and terroir employed by wine producers throughout the region and attempts to identify the expressed and implied causes of the relative failure of the wineries on the plains and peninsula to develop and prosper.

The Canterbury Wine Region

Canterbury is a province on the South Island of New Zealand. In total Canterbury accounts for only 3.9 percent of the country's vineyard area; by contrast New Zealand's major wine regions are Marlborough (67.7 percent) and Hawke's Bay (12.6 percent) (New Zealand Winegrowers 2017b, ii). In the country as a whole, the wine industry has undergone rapid growth and transformation since the mid-1980s. Although the industry had its earlier focus on the production of cheaper sweet wines almost exclusively for the domestic market, it has become much more export oriented and has grown steadily (the area under grapevines in the country has grown from a little more than five thousand hectares in 1990 to more than thirty-five thousand in 2016). Having moved significantly away from bulk wine production, the industry now produces wines that command among the highest per bottle prices on global markets (New Zealand Winegrowers 2017a). Alongside this growth has been a significant geographical reorientation. From earlier centers on the North Island (Auckland, Gisborne, and Hawke's Bay), the industry shifted south as it developed, and now the province of Marlborough in the north of the South Island is by far the largest producing region, accounting for more than two-thirds of the country's vineyard area and nearly three-quarters of its production (Moran 2016). Although there

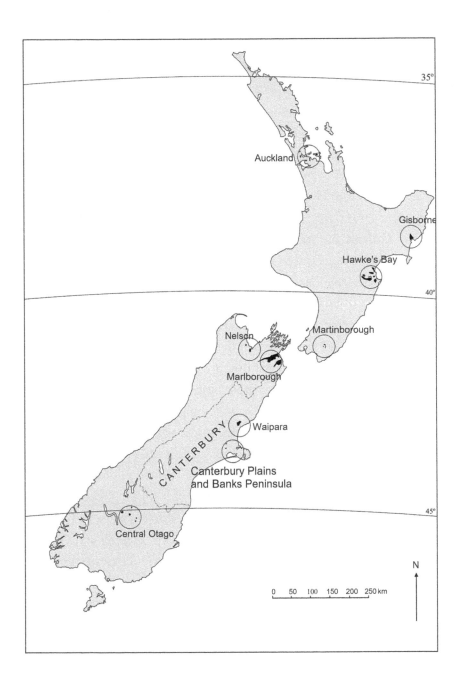

FIG. 7. New Zealand wine regions. Map drawn by John Overton (vineyard areas approximate).

has been this regional concentration—alongside a varietal specialization (on sauvignon blanc), several new wine regions have grown from almost nothing in the span of a few decades. Martinborough, Nelson, Waipara, Central Otago, and subregions elsewhere (Gimblett Gravels, Te Awanga, Waiheke Island) now feature as denominations of origin in New Zealand, even though wine production was virtually unheard of there before 1980 (Baragwanath and Lewis 2014; Cull 2001; Cooper 2008; Moran 2016; Howland 2014; Overton and Murray 2014a, 2014b). Canterbury was a part of this growth and spread of the wine industry in New Zealand, but its success has been rather uneven.

In 1991 the prospects were bright for the rapid development of a new wine region on the Canterbury Plains near the city of Christchurch. The region, and the St. Helena Winery in particular, had produced in 1982 what is regarded as the first example of an excellent pinot noir in the country (Stewart 2010, 348). In the following decade new vineyards were planted and wineries established, and there was a steady stream of good wines being produced: more pinot noir by St. Helena, riesling and chardonnay by Giesen, and a wide range of varietals by Torlesse. The rapid development of the Marlborough region, some four hours' drive to the north of Christchurch, provided an example of what could happen to a region when wine production developed a good reputation and generated momentum in its growth.

It seemed as if there were several key elements in place in Canterbury. First, the pioneering wines produced at St. Helena and Giesen had proven that the region had the potential to produce award-winning wines.[3] This would have seemed to suggest that the region possessed an environment suitable for quality grape production: well-drained gravels and thin alluvial soils in many places and a climate marked by rainfall, sunshine, and temperature regimes that suited cooler-climate varieties, such as pinot noir. There were even advantages to be had from the somewhat marginal climate: the Giesen winery found almost by chance that slow-ripening riesling grapes, threatened by rain and cooling temperatures, could develop botrytis and late-harvest characteristics helped by the föhn, a warm wind from the northwest, as occurred in 1989 and 1990.

Second, land was available. In the late 1980s, thanks to government deregulation and economic recession, land values had fallen throughout rural New Zealand and the traditional agricultural sectors of the economy—sheep (meat and wool) and cropping—were struggling. Pastoral blocks could be obtained and planted in grapes relatively easily, though capital was scarce. There was also technical advice in the form of the agricultural research conducted at Lincoln University, where a small cluster undertook experiments in grape growing and developed expertise they were keen to share with potential grape growers. Dr. David Jackson, a plant physiologist, and Daniel Schuster, a Czech consultant winemaker, were at the core of this group, and augmenting their efforts was a small collection of hobby grape growers and winemakers—most of them medical professionals (including neurologist Ivan Donaldson)—who had banded together and planted a range of grapes on a small plot outside Christchurch (Schuster, Jackson, and Tipples 2002).

Finally, and crucially, there was a ready-made local market nearby. Christchurch is the largest city on the South Island of New Zealand and had a population of about three hundred thousand in 1990. This urban center was a source of loyal consumers; wine consumption was rising steadily at the time, and consumers were turning from cheaper bulk wines to more distinctive varieties and locally produced wines. The city also provided potential investors and labor, as well as many who would seek out weekend visits to new wineries and country restaurants. With the success of St. Helena and Giesen (the latter winning a prestigious award in London for its riesling), there was a developing pride in Canterbury wines. Parochial support and some good vintages in the late 1980s and early 1990s spurred growth in several areas within reach of the city. There were wineries on the plains near Belfast, West Melton, and Burnham; the hilly volcanic landscapes of Banks Peninsula attracted attention, with a winery and restaurant at French Farm proving especially popular; and some growers began planting grapes

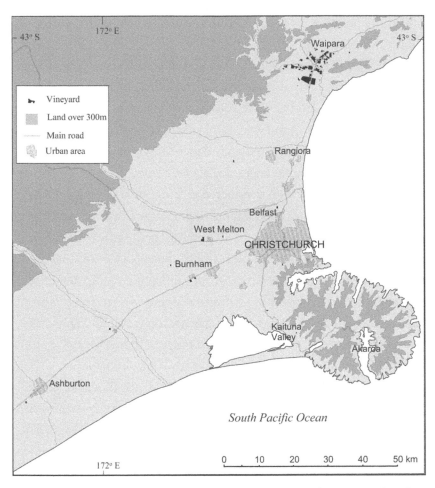

FIG. 8. Winegrowing areas in Canterbury. Map drawn by John Overton based on field observations.

with success near Waipara, which is a sixty-kilometer drive north of Christchurch (fig. 8).

Unfortunately, the optimism and expansion did not continue. The Lincoln University experiments had indicated that Canterbury experienced a relatively low number of degree-days over the October-to-April growing season (below the favored one-thousand mark) (Moran 2016, 336), and the vintages of 1992 and 1993 (probably affected by the climatic effects of the Mount Pinatubo eruption in 1991) were

FIG. 9. New vineyard development, Waipara, 2007. Photo by John Overton.

especially poor. Expansion virtually halted, and winemakers began to look elsewhere.

It was at this time that the developments at Waipara began to demonstrate that this subregion had a rather better climate for grapes: when winemakers compared notes, they found that the grapes ripened earlier and more reliably at Waipara than on the plains (Cooper 2008, 249). The early winemakers had included a joint venture that experimented successfully with chardonnay, riesling, and (much less so) cabernet sauvignon varieties. Pinot noir also did well. Significantly, the early riesling operation was sold to Corbans Wines—then one of the largest wineries in the country and soon itself to be bought out by the largest (Montana), which had most of its operations in Marlborough.[4] Montana decided to keep this holding in Waipara and indeed to expand by purchasing in 2002 a large new block on which to expand riesling and pinot noir production. There was also a very large development at Waipara, with the Waiata Vineyard covering three hundred hectares (fig. 9) and the grouping and expansion of a number of operations, including Canterbury House and Waipara Hills, under the Marlborough-based Mud House brand. Ivan Donaldson's family winery, Pegasus Bay, was also significant. Although smaller in scale than Montana, Waiata, or Mud House, it developed a winery and on-site restaurant, attracting visitors to the region and symbolizing the success and recognition of Waipara for fine wines and dining. Waipara has continued to expand while the Canterbury Plains and Banks Peninsula have languished.

Furthermore, the Waipara winemakers sought to establish themselves as an entity separate from the rest of Canterbury, forming the Waipara Winegrowers Association (later the Waipara Valley North Canterbury Winegrowers). This group tried to promote its wines as distinctive—perhaps with an eye to the less-than-spectacular reputation some wines from the rest of Canterbury were gaining. Starting in 2000, this association was able to forge a separate identity within the framework of the national industry body (New Zealand Winegrowers), having separate statistical summaries and recognition as a "wine growing area" (Overton, Banks, and Murray 2014, 248). The divergence of Waipara from the rest of Canterbury can also be seen in returns for vineyard area compiled by New Zealand Winegrowers (table 1).

As Waipara grew, the early wine operations of the Canterbury Plains stagnated. However, there were some notable departures. The Giesen winery at Burnham struggled after 1990 but had ambitions to expand. The company purchased its first property in Marlborough in 1993, concentrating on sauvignon blanc production, then opened a large winery in 2000, basically signaling a departure from Canterbury. Giesen has become one of the country's leading medium-scale producers and wine exporters. Two of the early leading winemakers in Canterbury were Daniel Schuster and Mark Rattray, both of whom made award-winning wines for St. Helena. Both decided to start their own vineyards—but in Waipara rather than on the plains—and they continued to build the reputation for pinot noir. Others making the move from the plains to Waipara included the Torlesse and Sherwood wineries (Tipples 2007).

TABLE 1. Growth of vineyard area planted in Waipara versus Canterbury, 1992–2016 (ha)

	WAIPARA	REST OF CANTERBURY	TOTAL CANTERBURY
1992			161
1994			208
1996			213
1998			350
2000	210	232	442
2002	248	234	482
2004	383	258	641
2006	652	272	924
*			
2012	1,034	163	1,197
2014	1,255	193	1,448
2016	1,251	168	1,419

Sources: New Zealand Winegrowers 2012; New Zealand Winegrowers (previously NZ Wine Institute) annual reports and statistical returns for 1992–2017.

* Data between 2007 and 2011 were not collected or are regarded as suspect (see New Zealand Winegrowers 2012, 3) and are not included.

The flight of capital and winemaking expertise, as well as the diverging growth of Waipara, left behind a wine landscape in the rest of Canterbury that was characterized by decline and sometimes neglect. Some vineyards, as in West Melton, were basically abandoned, others became contract growers selling grapes to winemakers elsewhere, and for a number of years the Giesen brothers struggled to sell their Burnham property. Things were little better on Banks Peninsula. Despite some success for the small Kaituna Valley Wines operation, the French Farm Winery faltered and became a restaurant-based operation.[5] However, not all was failure and decline. Rossendale Winery had been established on the outskirts of Christchurch, and its restaurant business and associated vineyard did well after 1994 (Moran 2016, 341–42), even though its owners also looked more to Marlborough to expand their overall winemaking business. Elsewhere there are less visible producers growing grapes on contract for Pegasus Bay, for example, or maintaining smaller family-run businesses (as at Kaituna Valley).

Today the two wine regions occupy separate spaces in the market and are very different in terms of scale and landscapes. Although Waipara lies within the broader Canterbury territorial region, it has its own geographical indication.[6] On the other hand, the geographical term "Canterbury" is used on labels of wines from both the plains and Banks Peninsula. Waipara has a more or less contiguous area of vineyards and the feel of a distinct region, albeit with some distinctions between the flats to the south and middle and the limestone hills to the east and foothills to the west. The hills to east and west frame the valley and provide protection from cooling winds so that much of the valley, when not irrigated, appears dry and grassy. There is a small town at the center, and vineyards are a prominent feature of a seemingly quite prosperous small rural district. In the vineyards themselves there is some varietal specialization with sauvignon blanc and pinot noir (and also riesling and pinot gris); the proportion of riesling is the highest among New Zealand wine regions (table 2). Farther south, however, the Canterbury wine region is virtually invisible. One has to search out vineyards across a wide area. There are a few winery/restaurant operations on the outskirts of the city; some small vineyards, usually without cellar door sales, and some with visible signs of neglect; and one or two small wineries tucked in sheltered valleys on Banks Peninsula. Interestingly, though, and perhaps reflecting the contracting arrangements with Waipara and other wineries, the composition of the vineyards is similar to that found in Waipara (see table 2). Yet, from a situation in which these two regions had a similar area under grapes in 2000, by 2016 the scattered Canterbury vineyards accounted for less than one-eighth of the Waipara area (see table 1).

TABLE 2. Composition of Waipara and Canterbury vineyards by area (ha) and percentage

| | WAIPARA | | CANTERBURY | |
VARIETY	(HA)	PERCENTAGE	(HA)	PERCENTAGE
Chardonnay	67	5.4	14	8.3
Gewurztraminer	16	1.3	2	1.2
Pinot gris	177	14.1	19	11.3
Pinot noir	339	27.1	65	38.7
Riesling	261	20.9	23	13.7
Sauvignon blanc	355	28.4	41	24.4
Other varieties	36	2.9	4	2.4
Total	1,251	100.0	168	100.0

Source: New Zealand Winegrowers (2017b).

Fermenting Failure

When comparing the story of these two small regions, it is tempting to see the failure of one and the success of the other in purely environmental terms. And that is the impression gained from the few accounts that have appeared (e.g., Tipples 2007; Moran 2016). The benign climate of Waipara, with its "sheltered" and "warm" features, complemented by limestone and rolling topography (even if most of the grapes are grown on flat plains with few or no traces of limestone), is contrasted with the fickle, exposed, and cold plains farther south.

There is certainly much to support this argument for the environment's role in determining winemaking success. Higher degree days in Waipara produce ripe grapes earlier more reliably in nearly every year. By contrast, grape growers nearer Christchurch face generally cooler temperatures during ripening, they have to deal with the strong winds that sweep unimpeded across the plains, and frosts and drought provide additional hazards. The factors were certainly at play with the poor harvests in the early to mid-1990s, which produced challenges that persuaded many to leave the region and reestablish themselves elsewhere. Some years may produce good harvests, but others will fail. That sort of pattern is simply too unreliable a basis on which to invest in winemaking in the region. There is thus a fairly clear argument that the Canterbury Plains and Banks Peninsula are simply not well suited to producing consistently good-quality grapes to make wines that can sustain a healthy wine industry.

However, before we accept this argument in its entirety, it is important to examine other factors at play, because areas such as the Canterbury Plains or Banks Peninsula, though climatically marginal for grape growing, might survive if the quality and reputation gained from good years—with occasional exceptional wines—could cover the poor ones.

First, Canterbury (other than Waipara) failed to establish itself in the public imagination as a wine region, and this had much to do with its landscape and topography. Waipara lies in a valley with defining ridgelines, more or less contiguous vineyards lining the valley floors and lower hillsides, and it has a definable center (even though this is merely a road junction and a small township). It feels like a wine region because of the density of vineyards and the sense of a distinct place separate from others. It is a destination. Visitors can feel they are entering the wine district, they can visit restaurants and buy wines from cellar doors at several locations, and the dry and grass-covered hillsides seem to contrast with and complement the neat green rows of vines. By contrast, the vineyards and wineries of the Canterbury Plains are scattered over a large area, and rarely does one vineyard neighbor another. Some lie on the direct outskirts of the city of Christchurch; others spread out in different directions, some many miles from the city. There is no center, no definable start and finish to the region. If one were to embark on a wine trail in Canterbury in 1990 (or today), most of one's time would be spent traveling by car from one remote vineyard to another. Small wineries and vineyards are separated by other dominant rural land uses: pastoral farms, cropping, forestry, and diverse uses such as quarries, horse riding schools, orchards, and hobby farms (known as lifestyle blocks).

The Canterbury Plains also defy delimitation. They are framed pictur-esquely by the distant Southern Alps to the west and Banks Peninsula to the east, but locations such as Burnham or West Melton are largely featureless points on a large, flat plain.

Elsewhere I have discussed how the establishment and success of "met-ropolitan wine regions" requires two different elements: functional prox-imity and existential separation (Overton 2017). Functional proximity involves the linking of urban capital, labor, and demand with nearby rural land and production and a sense of shared identity. Existential sep-aration, on the other hand, is more about the sense of being somewhere different and distinct. The wineries of the Canterbury Plains certainly possess functional proximity with investment and labor, as well as a potentially loyal parochial customer base in the city of Christchurch. But, unlike Waipara, existential separation is not there. Some wineries abut the suburbs; others are dotted among a mixed rural landscape. There is little sense of leaving the city and entering a different rural area dominated by vineyards.

The few vineyards on Banks Peninsula, however, do offer more poten-tial for a sense of place, identity, and existential separation. The road from Christchurch to the town of Akaroa—a popular weekend and recreation destination—winds through and over the peninsula, with many valleys along the way, each with a different aspect, with hills and established farms. The bucolic landscape would be a suitable place in which to see an idyllic wine region. Yet a wine region nevertheless failed to develop here, though for different reasons. The cool and variable climate on the peninsula means that only a few small sheltered locations—such as parts of Kaituna Valley—are able to sustain grape growing. Even a small wine district with several properties in proximity is not viable. Environmen-tal limitations on the peninsula (and the plains) thus combined with a weak sense of place and identity on the plains to limit the possibility for stirring the imagination of a favored wine region.

A second and related factor in the inability of Canterbury to emerge and develop successfully as a wine region was the absence of a terroir narrative; as suggested earlier, such narratives have been critical in the development of wine regions around the world. Winemakers and businesses have often proved adept at constructing stories that link their region to claims of uniqueness and quality linked to environmental factors, such as soils and climate, or historical traditions and narratives of experimentation and triumph. For Waipara, the winegrowers' association was able to do this with some ease. The landscape was suitable: the Teviot Hills to the east provide shelter from cooling easterly winds, there is some protection from damaging northwesterly winds, the valley alignment seems to afford maximum sun, and summer temperatures are higher than in surrounding areas. Furthermore, the environmental narrative can claim the advantages of underlying limestone geology (even though the bulk of the vineyard area is on the alluvial gravels to the south of the Waipara River, rather than on the lower slopes of the limestone hills). There are even some advantages claimed for the topography: many slopes face north or west and offer good cold-air drainage. Added to these environmental narratives in Waipara are some minor cultural ones, principally relating to the supposed almost heroic perseverance of pioneers such as John McCaskey, who proved the worth of the region.

For the Canterbury Plains, however, the terroir discourse has been much more problematic. Climate has been portrayed as a problem, not a bonus, being cooler and variable, with hazards from high winds, frost, and unwelcome southerly storms. The exposed nature of the plains and the peninsula makes these narratives hard to counter. Soils are also more difficult to portray positively, for the underlying gravels of the plains are largely undifferentiated and originate from the distant mountains. Aeolian deposits on the hillsides of the peninsula and its volcanic origins may offer some basis for terroir claims, but these have not been forthcoming. Although we have not seen a terroir narrative emerge, there may have been some foundation for one or more such narratives in the history of winemaking in Canterbury. One possibility lies in the story of how the medical professionals who experimented with different varieties in the 1970s found what grew well; some, such as Ivan Donaldson, eventually established their own vineyards and wineries. This would have made a coherent story of how local people, perhaps a little eccentric, found wines that could gain recognition and acclaim.

The story of Canterbury wine has been recounted in part by Daniel Schuster et al. (2002) but not disseminated or promoted by winemakers. Nor has another possibility: the claim that Canterbury can make to be the birthplace of the New Zealand pinot noir wine industry. That variety is now the country's second most planted variety and a very successful product and export for many regions. But it was the 1982 St. Helena pinot noir that was the first to gain a gold medal in a national competition, and it has been seen as first proof of the variety's potential in New Zealand. This wine too has an interesting backstory: although produced by the St. Helena Winery in Belfast, its grapes came from a small plot on the peninsula (what became the Kaituna Valley Wines operation), and it was made first by Daniel Schuster, thanks to his European experience. Here are the makings of a fine story to be exploited: a tiny patch of grapes planted in a small plot in an idyllic valley, a quirky winemaker from the Old World, and an old, established New Zealand farming family who experimented with wine to diversify their farming operations.

However, the terroir stories did not emerge in Canterbury. When the region needed these promotional devices most, in the hard years of the 1990s, there were few examples in the country, except perhaps Martinborough (Howland 2014), where they had been employed. In subsequent years areas such as the Gimblett Gravels, Central Otago, and Waiheke would, along with Martinborough, use terroir effectively to define and promote their regions (Murray and Overton 2011).

One reason for the lack of promotion—and a reason for the faltering development of Canterbury wine—was the absence of a strong industrial cluster. Whereas Waipara producers acted in concert in the late 1990s to form an association and promote their region as separate from Canterbury, the remaining Canterbury wineries appeared to be much more disjointed and without clear leadership. Again, geography was probably a factor. Winemaking operations were not close to one another, being scattered over a wide area. They also did not have a lot of common characteristics, beyond being small and struggling. It was hard for the winery/restaurant at French Farm on the Akaroa harbor (trying to establish an image of European connections and inspiration on a scenic hillside) to see itself as similar to the former potato farm at Belfast or the Giesen operation on the plains at Burnham. The disparate nature of the producers and the weakness of the wine cluster were not helped by the absence of larger players or a presence on the national market. Waipara had the Corbans/Montana connection, and this helped provide higher volumes of output that meant the Waipara denomination was seen on retail shelves throughout the country and sometimes overseas. In Canterbury the Giesen winery was soon among the largest, but its struggles soon led to its reorientation to Marlborough. After that it was not able to promote the region with its wines as it had done in 1989–90.

A final possible factor in Canterbury's failure to thrive relates to water. Although the Canterbury Plains are seen to be cooler than regions such as Marlborough or Waipara, the region suffers equally from occasional drought; without irrigation, grape yields are lower during droughts, which ultimately kill the vines. Most wine regions in New Zealand require irrigation to counter the effects of seasonal water shortage and maintain optimum water and sugar levels, leading to better economic yields. Waipara was fortunate in that an irrigation scheme, initially aimed at providing water for cropping and livestock in a district where drought had been a major hazard for farmers, had been built with government assistance in the 1970s. It proved to be a significant asset for grape growers and meant that yields could be maintained even in the driest of years. On the peninsula and plains, however, the areas where grapes were planted rarely had access to irrigation. There were some small local water schemes, as at West Melton, but most grape growers faced difficulties gaining access to water.

Thus, in comparing the success of the Waipara wine region with the failure of the rest of Canterbury to develop, we see a multitude of factors at work. Climatic differences were crucial. Lack of warmth over the critical grape growing and ripening seasons remains the key reason why the early vineyards on the Canterbury Plains and Banks Peninsula failed to thrive and develop into a sustainable and prosperous wine region. Yet this was not the only reason, and other advantages present—particularly the presence of a solid local customer base—might have compensated for this climatic disadvantage. And in Waipara favorable climate alone was no guarantee of success.

To use the metaphor of fermentation, for a wine region to develop there must be several key ingredients present, and they must combine—and "ferment"—to produce a metamorphosis in the landscape. This involves the bringing together of key factors (environment, capital, know-how, leadership, cooperation, etc.); it requires time for relationships to develop and for learning processes to take place; and it frequently involves a random aspect (perhaps analogous to the presence of wild yeasts) whereby luck or misfortune, fortuitous or bad timing, can tilt outcomes one way or the other. Waipara has a combination of suitable climate and soils that make grape growing possible. It also has the ingredients to turn these factors into a terroir narrative: climate, soils, aspect, topography, history, and a collection of skilled winemakers. It has an industrial cluster in terms of an effective winegrowers association and obvious cooperation and synergies among the winemaking operations (see chapter 2, this volume). This cluster was helped by the involvement of large wine companies prepared to invest, develop new vineyards, and use the Waipara name. It has water. It has proximity to a metropolitan center, yet it is just far enough from Christchurch for people to feel that they are going to a destination that is viscerally different from the city. It has an identity in terms of landscape and boundaries.

On the other hand, the plains of Canterbury and the valleys and slopes of Banks Peninsula could muster only a few of these ingredients. Its identity was weak and fragmented, there was little leadership and cooperation to define and promote it, and it suffered both from being arguably too close to Christchurch yet also too distant from it. It was not a destination; it was not a "region" in the imagination of customers.

Sometimes the right factors came together and good wines were produced in some years, but mostly it has languished or at best stagnated. It is a wine landscape that has failed to ferment.

Conclusions

What then might we make of this New Zealand example in considering the way wine landscapes evolve and become established? First, success is not inevitable. Despite the proliferation of small-scale winegrowing in many parts of the world and the growth of wine regions away from the established wine industrial centers, grape growing and winemaking remains a difficult proposition. It requires land and capital, and bottles of wine can be hard to sell in a crowded and highly competitive global market. There is clearly a place for smaller regions to develop a loyal parochial customer base—as with metropolitan wine regions—but the proximity of a large urban population of wine consumers does not guarantee the success of nearby wine-producing areas. Perhaps academic analyses of the development of new wine regions have been just as affected by the romantic visions of idyllic vineyard landscapes and artisanal winemaking as have been the ambitions of many erstwhile small-scale winemakers. Risk, struggle, debt, setbacks, and failure are as much a part of the narratives of winemaking as are awards, quality, popularity, and success. And while many new wine regions have become established, many others experience a precarious and faltering existence and even failure.

Second, we should be wary of relying on the sort of environmental determinism that terroir discourses suggest as the key explanation for the success of wine regions. A suitable climate for the growing and ripening of grapes is clearly a necessary condition for a wine region. And soils, underlying geology, topography, and aspect may all contribute to notions of supposed quality. This study exposes how two regions (Waipara and Canterbury), despite being close to each other, developed quite differently. One has a demonstrably better climate for grapes, and it has tended to develop at the expense of the other. However, these factors alone provide a poor explanation for the success or failure of wine regions. Wine can be produced from environments that would not rank highly in any objective analysis of climatic suitability. The Champagne region of France, for example, is hardly a perfect grape-growing climatic area, with noted temperature and sunshine deficits, yet it is highly successful as a result of historical circumstance and clever marketing. Elsewhere we see grape-growing and wine regions becoming established in places that would defy scientific justification in terms of climate: a Polynesian atoll, Thailand, Tanzania, Hokkaido in Japan, and so on (Banks et al. 2013, for example). Indeed climatic marginality can be seen as a distinct advantage for winegrowing in some cases: poor harvests in some years can be offset by some vintages where a degree of vine stress produces grapes of deep flavor intensity and wines of great quality. So climate and soils—and the range of other terroir-based components—do not by themselves explain why some wine landscapes develop and others do not.

Instead we need to realize that wine regions are constructed and landscapes ferment. Critical in this process is the way public imaginaries are created and manipulated. Important here is the natural landscape. Hills can frame and delimit an area so that it feels different and distinctive, and many wine consumers associate the wine they buy and consume with an environmental imaginary (Taylor and Hurley 2016)—or experience—of the place it comes from. A pretty valley or a scenic hillside cloaked in vineyards is much more appealing (perhaps regardless of its summer heat balance or frost incidence) in the minds of wine buyers than a flat, featureless expanse of diverse rural activities in which vineyards appear rarely and randomly. The idea and perception of a region are critical, and this is where industrial contiguity, cooperation, and promotion become important (Brémond 2014; Charters and Michaux 2014). Where vineyards cluster together, where they promote themselves collectively, and when they inspire a vision of bucolic lifestyles and products of quality, character, and uniqueness (Getz and Brown 2006), then wine regions will be made as much by deliberate human agency and storytelling as by supposed environmental advantages. In this we can see the importance of collective action and industrial clustering. Finally, however, we also need to appreciate a certain random element: the ingredients may all be in place, but whether or not they successfully ferment to produce a desired reaction and a successful wine landscape can be affected by chance. A succession of poor vintages, inadequate leadership, changing consumer tastes, or natural disasters can mean that the various components do not combine effectively, while on the other hand an unexpected trophy award, inspired and cohesive collective action, or an exceptional vintage can provide catalysts for success. "Fermented landscapes" are not just reflections of the products they produce; they are themselves the result of a complex alchemy.

Notes

1. There are links here with work on the emerging concept of "exurbia" (Taylor and Hurley 2016; Hiner 2016), or places where the urban and rural are intermingled and fused.
2. The Canterbury Region is defined in figure 7 by the boundaries of the Canterbury Regional Council (ECan).
3. Interestingly, these two pioneering wineries and brands have experienced quite differing fortunes. Despite its early successes, St. Helena Winery was advertised in a liquidation sale in 2011 and now appears to be out of business. However, as described later in this chapter, Giesen Wines has moved to Marlborough and become a successful and established sauvignon blanc producer and exporter.
4. Montana was purchased later by the global beverage corporation Pernod Ricard, and the Montana brand has been virtually retired.
5. Interestingly, the Kaituna Valley winery's grapes—not those from its Belfast property on the Canterbury Plains—had been used to make the breakthrough 1982 St. Helena pinot noir.
6. New Zealand does not yet have a fully functioning geographical indications regulatory framework, in the manner of the AVA of the United States or France's AOC, despite the passing of enabling legislation. However, other mechanisms, such as trademarks and the oversight of New Zealand Winegrowers, have been used to established various "place brands" of New Zealand wines, and agreements with the European Union have codified various protected geographical indications for New Zealand wine in overseas markets (Overton and Murray 2017).

References

Banks, Glenn, Ratchaphong Klinsrisuk, Sittipong Dilokwanich, and Polly Stupples. 2013. "Wines without Latitude: Global and Local Forces and the Geography of the Thai Wine Industry." *EchoGéo* (online) 23 (1). https://doi.org/10.4000/echogeo .13368.

Banks, Glenn, and John Overton. 2010. "Old World, New World, Third World: Reconceptualising the Worlds of Wine." *Journal of Wine Research* 21 (1): 57–75.

Banks, Glenn, and Scott Sharpe. 2006. "Wines, Regions and Geographic Imperative: The Coonawarra Example." *New Zealand Geographer* 62 (3): 173–84.

Baragwanath, Lucy, and Nicolas Lewis. 2014. "Waiheke Island." In *Social, Cultural and Economic Impacts of Wine in New Zealand*, edited by Peter J. Howland, 211–26. Abingdon: Routledge.

Barham, Elizabeth. 2003. "Translating Terroir: The Global Challenge of French AOC Labelling." *Journal of Rural Studies* 19 (1): 127–38.

Bramble, Linda, Carman Cullen, Joseph Kushner, and Gary Pickering. 2007. "The Development and Economic Impact of the Wine Industry in Ontario, Canada." In *Wine, Society and Globalization: Multidisciplinary Perspectives on the Wine Industry*, edited by Gwyn Campbell and Nathalie Guibert, 63–86. New York: Palgrave Macmillan.

Brémond, Joël. 2014. "Rioja: A Specific and Efficient Economic Model for Wine Region Organization." *Journal of Wine Research* 25 (1): 19–31.

Charters, Stephen. 2006. *Wine and Society: The Social and Cultural Context of a Drink*. Oxford: Elsevier.

Charters, Steve, and Valery Michaux. 2014. "Strategies for Wine Territories and Clusters: Why Focus on Territorial Governance and Territorial Branding?" *Journal of Wine Research* 25 (1): 1–4.

Cooper, Michael. 2008. *Wine Atlas of New Zealand*. Auckland: Hodder Moa Beckett.

Cull, Dave. 2001. *Vineyards on the Edge: The Story of Central Otago Wine*. Dunedin, New Zealand: Longacre Press.

Demossier, Marion. 2004. "Contemporary Lifestyles: The Case of Wine." In *Culinary Taste: Consumer Behaviour in the International Restaurant Sector*, edited by Donald Sloan, 93–107. Oxford: Elsevier Butterworth-Heinemann.

———. 2005. "Consuming Wine in France: The 'Wandering' Drinker and the *Vinanomie*." In *Drinking Cultures: Alcohol and Identity*, edited by Thomas M. Wilson, 129–53. New York: Berg.

Fanet, Jacques. 2004. *Great Wine Terroirs*. Berkeley: University of California Press.

Fløysand, Arnt, Stig-Erik Jakobsen, and Ove Bjarnar. 2012. "The Dynamism of Clustering: Interweaving Material and Discursive Processes." *Geoforum* 43 (5): 948–58.

Getz, Donald, and Graham Brown. 2006. "Critical Success Factors for Wine Tourism Regions: A Demand Analysis." *Tourism Management* 27 (1): 146–58.

Giuliani, Elisa, Carlo Pietrobelli, and Roberta Rabellotti. 2005. "Upgrading in Global Value Chains: Lessons from Latin American Clusters." *World Development* 33 (4): 549–73.

Hiner, Colleen C. 2016. "Divergent Perspectives and Contested Ecologies: Challenges in Environmental Management along the Rural-Urban Interface." In *A Comparative Political Ecology of Exurbia: Planning, Environmental Management, and Landscape Change*, edited by Laura E. Taylor and Patrick T. Hurley, 51–82. Cham, Switzerland: Springer International.

Howland, Peter. 2014. "Martinborough: A Tourist Idyll." In *Social, Cultural and Economic Impacts of Wine in New Zealand*, edited by Peter J. Howland, 227–42. Abingdon: Routledge.

Moran, Warren. 2001. "Terroir—The Human Factor." *Australian and New Zealand Wine Industry Journal* 16 (2): 32–51.

———. 2016. *New Zealand Wine: The Land, the Vines, the People*. Auckland: Auckland University Press.

Morris, Arthur. 2000. "Globalisation and Regional Differentiation: The Mendoza Wine Region." *Journal of Wine Research* 11 (2): 145–53.

Murray, Warwick E., and John Overton. 2011. "Defining Regions: The Making of Places in the New Zealand Wine Industry." *Australian Geographer* 42 (4): 419–33.

New Zealand Winegrowers. 2012. *Vineyard Register Report 2012*. Auckland: New Zealand Winegrowers.

———. 2017a. *Annual Report 2017*. Auckland: New Zealand Winegrowers.

———. 2017b. *Vineyard Register Report 2016–2019*. Auckland: New Zealand Winegrowers.

Overton, John. 1996. "The Wine Industry." In *Changing Places: New Zealand in the Nineties*, edited by Richard B. Le Heron and Eric J. Pawson, 150–54. Auckland: Longman Paul.

———. 2017. "Rural Idylls and Urban Economies: Cities and the Development of Wine Regions." Conference address, Wine and Culinary Tourism Futures Conference, Kelowna BC, Canada, October 20, 2017.

Overton, John, Glenn A. Banks, and Warwick E. Murray. 2014. "Waipara." In *Social, Cultural and Economic Impacts of Wine in New Zealand*, edited by Peter J. Howland, 243–52. Abingdon: Routledge.

Overton, John, and Jo Heitger. 2008. "Maps, Markets and Merlot: The Making of an Antipodean Regional Wine Appellation." *Journal of Rural Studies* 24 (4): 440–49.

Overton, John, and Warwick E. Murray. 2014a. "Boutiques and Behemoths: The Transformation of the New Zealand Wine Industry 1990–2012." In *Social, Cultural and Economic Impacts of Wine in New Zealand*, edited by Peter J. Howland, 25–40. Abingdon: Routledge.

———. 2014b. "Finding a Place for New Zealand Wine: *Terroir* and Regional Denominations." In *Social, Cultural and Economic Impacts of Wine in New Zealand*, edited by Peter J. Howland, 41–57. Abingdon: Routledge.

———. 2016. "Fictive Place." *Progress in Human Geography* 40 (6): 794–809.

———. 2017. "GI Blues: Geographical Indications and Wine in New Zealand." In *The Importance of Place: Geographical Indications as a Tool for Local and Regional Development*, edited by William van Caenegem and Jen Cleary, 197–220. Cham, Switzerland: Springer.

Overton, John, Warwick E. Murray, and Fernando Pino Silva. 2012. "The Remaking of Casablanca: The Sources and Impacts of Rapid Local Transformation in Chile's Wine Industry." *Journal of Wine Research* 23 (1): 47–59.

Sánchez-Hernández, José Luis, Javier Aparicio-Amador, and José Luis Alonso-Santos. 2010. "The Shift between Worlds of Production as an Innovative Process in the Wine Industry in Castile and Leon (Spain)." *Geoforum* 41 (3): 469–78.

Schuster, Danny, David Jackson, and Rupert Tipples. 2002. *Canterbury Grapes and Wines 1840–2002*. Christchurch: Shoal Bay Press.

Sommers, B. J. 2008. *The Geography of Wine: How Landscapes, Cultures, Terroir, and the Weather Make a Good Drop*. New York: Plume.

Stewart, Keith. 2010. *Chancers and Visionaries: A History of Wine in New Zealand*. Auckland: Godwit.

Swart, Elmari, and Izak Smit. 2009. *The Essential Guide to South African Wines*. 2nd ed. Green Point, South Africa: Cheviot.

Taylor, Laura E., and Patrick T. Hurley, eds. 2016. *A Comparative Political Ecology of Exurbia: Planning, Environmental Management, and Landscape Change*. Cham, Switzerland: Springer.

Thomson, Joëlle. 2012. *The Wild Bunch: Movers, Shakers and Groundbreakers of the New Zealand Wine Industry*. Auckland: New Holland.

Tipples, Rupert. 2007. "Wines of the Farthest Promised Land from Waipara, Canterbury, New Zealand." In *Wine, Society and Globalization: Multidisciplinary Perspectives on the Wine Industry*, edited by Gwyn Campbell and Nathalie Guibert, 241–54. New York: Palgrave Macmillan.

Ulin, Robert. C. 1995. "Invention and Representation as Cultural Capital: Southwest French Winegrowing History." *American Anthropologist* 97 (3): 519–27.

———. 2007. "Writing about Wine: The Uses of Nature and History in the Wine-Growing Regions of Southwest France and America." In *Wine, Society and Globalization: Multidisciplinary Perspectives on the Wine Industry*, edited by Gwyn Campbell and Nathalie Guibert, 43–62. New York: Palgrave Macmillan.

Wilson, James E. 1998. *Terroir: The Role of Geology, Climate, and Culture in the Making of French Wine*. London: Mitchell Beazley.

Part 2

Landscapes of Ferment,
Alcoholic or Otherwise

Leaving the Old Kentucky Home   4

Emerging Landscapes of Bourbon Production

Christopher R. Holtkamp, Brendan L. Lavy,
and Russell C. Weaver

Bourbon whiskey is a distinctly American product. Historically, the majority of bourbon has come from distilleries in Kentucky, where its origins are tightly entwined with the state's physical and cultural landscapes. However, with the exploding popularity of bourbon internationally, craft distillers are opening across the United States, pushing bourbon production away from its physical and cultural heartland into new territories as diverse as the desert Southwest and downtown Brooklyn. In this chapter we examine the physical and cultural landscapes of newly established craft bourbon distilleries.

First, we identify non-Kentucky bourbon distilleries in the United States and document some of their physical landscape characteristics, including local geology and agricultural production, particularly corn, and compare these characteristics to the Kentucky areas that have historically produced America's bourbon. Next, we detail the cultural landscapes of new bourbon distilleries—as documented in their online marketing materials—to understand how new producers market bourbon in areas without the heritage and history that have been a cornerstone of bourbon identity for so long. Along those lines, we answer the following questions: To what extent do newer producers locate their distilleries in environments similar to Kentucky? How are new bourbon producers creating the authenticity and heritage that bourbon consumers expect? How are these producers, who do not have the pedigree of decades of production, driving business and building interest in their product?

A Brief History of Bourbon

Before discussing the history of bourbon, it is important to understand what makes bourbon distinct from other whiskeys. Bourbon is defined by the federal government as being produced "within the borders of the United States (not just Kentucky); it must be at least 51 percent corn, and it has to be aged in charred new oak barrels" (Mitenbuler 2015, 15). These standards were recognized by Congress in a 1964 proclamation making bourbon "America's whiskey." It is the corn that makes bourbon distinctive; in Europe most whiskey is made from barley and rye. In the New World corn proved to be a more successful crop, and its potential for distillation was first explored by George Thorpe, who was credited with the first distillation of corn—in 1620 near the Jamestown colony (Mitenbuler 2015). Although his spirit was far from the smooth bourbon consumed today, it marked one of the first efforts to utilize corn for whiskey production.

The bourbon industry, however, adopted and still tells a different origin story. Specifically, the industry tends to recount the tale of Rev. Elijah Craig (a name still used on a brand of bourbon produced by Heaven Hill Distillery), who distilled corn whiskey near Georgetown (in Bourbon County, Kentucky) in the late 1700s. The tale proceeds as follows. A barn fire damaged some oak barrels, but the thrifty pastor decided to use them anyway to store the corn whiskey he produced. He noticed the distinctive flavor and mellowness of the resulting whiskey, which inspired other distillers to use charred oak barrels for storing their whiskey (McKeithan 2012; Minnick 2016). Although this story is widely accepted, the reality is less romantic and far less marketable.

Efforts to transform corn into whiskey exploded as settlers moved west, particularly into the Kentucky region. Fertile soil, especially along the Mississippi Plateau and Bluegrass physiographic regions, provided for abundant corn production. Farmers turned to distillation as a means of utilizing excess production and monetizing their corn (McKeithan 2012). In the early 1800s corn-based spirits began to overtake the popularity of rye, which had been the primary grain used in American whiskey production (Minnick 2016). These early distillations did not produce the bourbon we know today. Still, through their efforts whiskey distillers were fine-tuning the methods for bourbon production while simultaneously creating a deep-rooted connection to Kentucky as the bourbon heartland.

Kentucky distillers used oak barrels to store and transport their whiskey and often loaded the barrels onto rafts for transport to New Orleans (Mitenbuler 2015). The journey often took months or longer, allowing the whiskey to age and incorporate flavors from the oak barrels. As early as the Roman era, it was known that charring the inside of a barrel would keep items fresh longer (Mitenbuler 2015). Charring may also have been used to sterilize barrels, with no intention of using the process to add flavor (Fryar 2009). French brandy producers used charred barrels to age their product, and French settlers, including the Tarascon brothers from the Cognac region, brought this tradition to Kentucky (Whiskey Advocate 2017). The earliest reference to this technique is from 1826, when a Lexington grocer requested whiskey stored in burned barrels (Minnick 2016).

In 1821 whiskey distillers Stout and Adams advertised their corn-based whiskey in the *Western Citizen Newspaper* as "'bourbon whiskey by barrel or keg'" (Minnick 2016, 15). This is the earliest known use of the term "bourbon." Although the name is most often associated with Bourbon County, Kentucky, scholars do not fully agree that the name is derived from the location. Several researchers associate the name with Bourbon Street in New Orleans, where corn-based whiskey from Kentucky was transported and sold (Mitenbuler 2015). Additionally, the House of Bourbon, a royal family of France, lent its name to both Bourbon County and Bourbon Street, and at that time the term "bourbon" was also adopted to denote the highest-quality products (e.g., bourbon coffee, sugar, and cotton were advertised as the best available) (Minnick

2016). Distillers most likely adopted the name "bourbon" for Kentucky corn whiskey both to denote its origin and to connect it to the quality associated with other "bourbon" products.

Although the bourbon industry embraces Elijah Craig and his tale as the creator of bourbon whiskey, there is stronger evidence pointing to a man named Jacob Spears as being the true founder of bourbon as we know it today (Minnick 2016). Around 1790 he established what is thought to be the first corn-based whiskey distillery in Bourbon County, Kentucky. Newspaper stories and family records support this claim. Additionally, Rep. Virgil Chapman, speaking in a 1935 congressional hearing, stated, "I do know that as an accurate, historical fact, in the year 1790, [two] years before Kentucky was admitted to statehood, a man by the name of Jacob Spears, in Bourbon County, Kentucky, where I reside now, made straight Bourbon whiskey, and because it was made in Bourbon County, that type of whiskey, wherever made in the world, has been called Bourbon whiskey ever since" (quoted in Minnick 2016, 33).

As bourbon production expanded, production techniques became more formalized and regulated over time. A lack of quality and consistency among whiskey producers led to the development of production standards to protect the image of bourbon (Mitenbuler 2015). To be labeled "Kentucky Straight Bourbon," the product had to be unadulterated whiskey aged for a minimum of two years (Minnick 2016). In 1897 Congress passed the Bottled-in-Bond Act that required any whiskey labeled as such to be produced at a single distillery, in one distilling season, aged at least four years, and bottled at 100 proof (i.e., 50 percent alcohol by volume) (Mitenbuler 2015). These early efforts protected the integrity and identity of bourbon as well as safeguarded public health from a number of similar but harmful products being sold at the turn of the twentieth century.[1]

In the post-Prohibition era bourbon production became much more concentrated. During this time a handful of distillers produced the vast majority of bourbon. These distillers leveraged the history and identity of bourbon, using names like Elijah Craig, Pappy Van Winkle, and other early distillers for their brands (Minnick 2016). Bourbon remained America's drink of choice through the 1960s until clear spirits became more popular, fueled in part by James Bond's preference for a martini, shaken, not stirred (Minnick 2016). As a result, bourbon fell out of favor, sales decreased, and production waned. Since the turn of the twenty-first century, however, bourbon has witnessed a resurgence, gained in popularity, and expanded in production (McKeithan 2012). Increasing popularity, coupled with states loosening liquor restrictions, has led to rapid growth in bourbon production, both from megadistillers and from the rapid surge in openings of craft distilleries in nearly every state of the Union. Bourbon, despite its close association with Kentucky, is truly becoming America's whiskey.

Physical Landscapes of Kentucky Bourbon

As mentioned above, bourbon production originated and long remained centered in Kentucky in large part because the region's physical land-scapes provided the necessary ingredients for bourbon production. In 1849 Kentucky was the second-largest corn producer in the nation after Ohio (Kemmerer 1949). The state's robust corn production is attributed in large part to the fertility of the rich, black soil in the region (Kem-merer 1949). Additionally, agricultural products from Kentucky had access to "cheap transportation over the Ohio and Mississippi Rivers" (Clark 1929, 539). Despite access to river transportation, growers still had difficulty getting corn to market in a timely fashion, especially because the country's population was largely east of the Appalachian Mountains and inaccessible via water routes. This led to corn being distilled into whiskey, as it represented "a more compact and portable 'value-added commodity'" that could be moved to market more easily (*New York Times Magazine* 2003). Corn brought ten cents per bushel, but whiskey sold for twenty-five cents per gallon, and a bushel of corn produced four to five gallons of whiskey (Kroll 1967). The added value, combined with the ease of transportation, made distillation an obvious choice to assure maximum revenue.

Water is a vital component in the production of bourbon, and many whiskey distillers and enthusiasts insist that the unique flavor of bourbon comes from the limestone-filtered water found in Kentucky wells and springs (Fryar 2009). There are two distinct geologic regions where bourbon is produced in the state: (1) the Inner Bluegrass and (2) Outer Bluegrass physiographic regions. The Inner Bluegrass region "is predominantly underlain by Ordovician limestone, while the Outer Bluegrass is underlain by Ordovician to Silurian limestone and shale and by Devonian limestone" (Fryar 2009, 607). This geology not only provides abundant groundwater for crops and the distilling process. Most importantly for bourbon production, the limestone also filters iron from the water; if iron were present in the water it "would make whiskey bitter and black" (Allen 1998, 87). To this day many Kentucky distilleries and some distilleries outside of Kentucky include a description of the water used for distillation in their marketing, indicating its importance to quality bourbon (fig. 10).

Cultural Landscapes of Kentucky Bourbon

Since its beginnings, bourbon production has fashioned distinct cultural landscapes that reflect its collective heritage and unique identity. In particular, early distillers and political conflict played an important role in the creation of bourbon's cultural landscapes, and contemporary bourbon distillers often employ the rich (at times exaggerated) stories in their production processes and marketing materials.

Bourbon's cultural landscape begins with its earliest European settlers in the region. Many early arrivals in Kentucky were Scots-Irish, and several of these settlers arrived with knowledge and skills of distillation—knowledge and skills that were eventually applied to the abundant corn that grew in the state's fertile soil (Mitenbuler 2015). The Whiskey Rebellion of 1794 led many settlers to flee the conflict that erupted in Pennsylvania. Some sought refuge in Kentucky, and many modern distillers identify the Whiskey Rebellion as their roots. However, "there were five hundred stills set up in Kentucky when it became a state (1792),

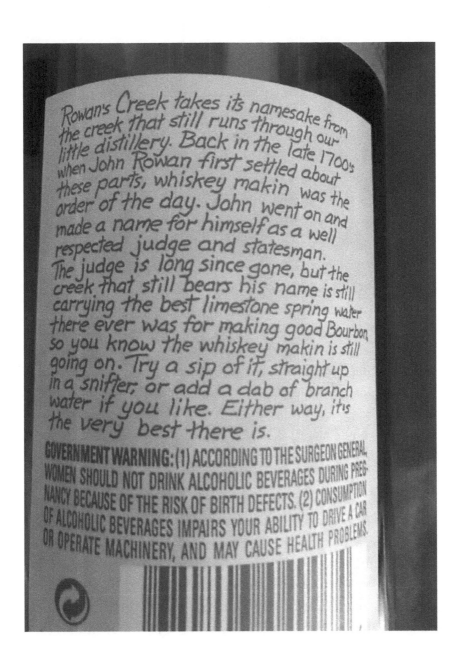

FIG. 10. Reference to water on bourbon label. Photo by Christopher R. Holtkamp, December 2017.

indicating that the Whiskey Rebellion likely had a minimal impact on the state's distilling industry" (Minnick 2016, 15). This effort, however, to connect to a historical event, even tenuously, is part of the bourbon industry's attempt to forge links to the heritage and identity of the past.

Bourbon distillation in Kentucky was (and remains) an extremely active industry, and many individuals made and sold whiskey prior to the Prohibition era. Over time the many independent distillers merged. Prohibition hastened consolidation, and after the end of that era, a few Kentucky-based megadistillers emerged as the center of bourbon production. Despite this, Kentucky bourbon distillers still cling to the identity and heritage of yeoman distillers and the rugged individualism of the frontier image (McKeithan 2012).

Contemporary bourbon brands like Elijah Craig, Evan Williams, Jim Beam, and Pappy Van Winkle hearken back to the early distillers of the region, yet they are now produced by international conglomerates (Minnick 2016). Despite bourbon being a multibillion-dollar industry, distilleries today recognize the value of historical identification and sell that heritage as part of the bourbon experience (Mitenbuler 2015). Figure 11 shows how a new distiller (Limestone Branch) makes an explicit connection to bourbon history through the adoption of the Yellowstone brand, which was initially produced in the 1870s to celebrate the creation of Yellowstone National Park. Bourbon identity has been defined by its history, and that connection to history has an influence on consumers' selection of a particular brand of bourbon.

Bourbon history and heritage are also connected to the modern identity of Kentucky. Bourbon production employs nine thousand workers and generates more than $125 million in tax revenue. The Bourbon Trail, established in 1999, attracts more than nine hundred thousand visitors annually from across the world (Kentucky Distillers' Association 2016). Distillery tours typically involve a presentation of how bourbon is produced, as well as the history of that specific distillery and how it connects to the past and the origins of bourbon. The challenge for the visitor is "deciphering the promotional from the factual materials presented by

FIG. 11. Bourbon branding often seeks a connection to history. The Yellowstone brand originated in the 1870s to celebrate the creation of Yellowstone National Park. Photo by Christopher R. Holtkamp, February 2018.

the proprietary establishment" (Simpson 2008, 89). Truth and fiction merge to create an identity "that prudently respects the promptings of the market and the choices of consumers, while also honoring the momentum of history by remembering that 'adaptation . . . is always better than invention'" (McClay and McAllister 2014, xi).

The discussion of how regional physical and cultural landscapes have shaped bourbon identity and in turn have themselves been shaped by bourbon production is important because it influences perceptions and marketing of bourbon today. Distillers celebrate the mythology of early distilling in the state—and the men credited with creating bourbon—as a uniquely American spirit. This distilled history overlooks the significant role of women and slaves, many of whom had active roles in managing farms and bourbon production (McKeithan 2012; Minnick 2016). The rugged character of the men who settled Kentucky, along with the bounty of the landscape, has shaped how companies have traditionally presented bourbon as the drink of men, allowing those who drink it to feel a connection to frontier heritage (McKeithan 2012). The focus on marketing to men, however, is beginning to change, as distillers recognize a missed opportunity and attempt to attract women to their products (Mitenbuler 2015).

The connection to the cultural and physical landscape of Kentucky is showcased in the internet presence of Kentucky distillers. The overwhelming majority of distilleries featured in searches included either place-names or other historic reference in their brands, such as Town Branch (a creek in Lexington, Kentucky, near the distillery) or Barton 1792, referencing the year Kentucky became a state. A more thorough discussion of our analysis is found in subsequent sections.

Purpose of the Study

Place characteristics, including history, are very important to Kentucky bourbon distillers and serve as avenues to entice consumers who connect to place through the consumption of bourbon. However, bourbon production has grown rapidly and diffused across the United States in recent years. In an effort to understand the recent growth and diffusion of bourbon distilleries, we document both the physical and cultural landscapes of new bourbon production centers across the United States and provide a geographic analysis of bourbon production and its local characteristics, including limestone geology and corn production, along with a content analysis of distillery websites to understand what relationship new distilleries have to the foundations of Kentucky bourbon.

Materials and Methods

We gathered data relative to physical landscapes and local geology from the National Agricultural Statistics Service of the U.S. Department of Agriculture (USDA) as well as the U.S. Geological Survey (USGS). We also acquired a list of bourbon distilleries through targeted internet searches. We primarily used a list curated by "Sku," an American spirits blogger (available at http://recenteats.blogspot.com/p/the-complete -list-of-american-whiskey.html). The list was current as of May 2017. It should not be considered an exhaustive list of bourbon distilleries; however, for the purposes of our research, it provided the names of 240 distilleries, at least one in nearly every state. We obtained data on bourbon distilleries and their marketing practices from their websites. Our final analysis included 242 distilleries across forty-five states and Washington DC.

To understand the cultural landscape of new bourbon production, we performed a content analysis of information contained in the websites of bourbon distilleries across the country. Content analysis is recognized "as a flexible method for analyzing text data" (Hsieh and Shannon 2005, 1277), and it provides a structure for creating "replicable and valid inferences from texts (or other meaningful matter) to the contexts of their use" (Krippendorf 2013, 24). In this case we used a directed content analysis approach with a coding structure derivative of our primary research goals (table 3). Our content analysis specifically examined how, if at all, bourbon producers outside of Kentucky use similar themes of heritage and history to market their bourbon.

TABLE 3. Cultural and physical landscape identifiers in bourbon marketing

Place-name	Bourbon/distillery contains a geographic or local place reference
Historic name	Bourbon/distillery contains a historic name or reference
History/heritage	Bourbon/distillery website references history and heritage (e.g., connection to historic distillers or place history)
Water	Bourbon/distillery references water used to make product
Local grains	Reference made to local grain used in product

Created by the authors with data from USDA Crop Frequency Layer, the USGS, and *Sku's Recent Eats*, http://recenteats.blogspot.com/p/the-complete-list-of-american-whiskey.html.

TABLE 4. Content analysis of distillers' marketing terms and descriptions

IDENTIFIER	KY COUNT	KY PERCENTAGE	NON-KY COUNT	NON-KY PERCENTAGE	TOTAL COUNT	TOTAL PERCENTAGE
Place-name	12	54.5	97	44.1	109	45.0
Historic name	11	50	52	23.6	63	26.0
History/ heritage	18	81.8	64	29.1	82	33.9
Water	9	40.1	41	18.6	50	20.7
Local grains	9	40.1	142	64.5	151	62.4

Created by the authors with data from USDA Crop Frequency Layer, the USGS, and *Sku's Recent Eats*, http://recenteats.blogspot.com/p/the-complete-list-of-american-whiskey.html.

Our coding scheme followed physical and cultural landscape terminology invoked by Kentucky distilleries and included identifiers common to these themes. We reviewed the websites of 242 distilleries to identify common language and descriptions used across the two primary categories of physical and cultural indicators. Specifically, we documented the occurrence of place-names, historic names, instances of history or heritage, and references to water and local grain use so that we could understand if and how new bourbon distillers connect the history and heritage of bourbon production at large and/or make specific linkages to their local communities. We also wanted to gauge connections to physical landscape characteristics of water and corn production, two key components of bourbon. Table 4 presents the results of our content analysis.

To gain a sense of the physical landscape, we geocoded distilleries to their cities of operation and ultimately situated them in their county subdivisions. For nearly all U.S. states, county subdivisions are "minor civil divisions" that align with the boundaries of boroughs, towns, and townships (Weaver and Holtkamp 2016, 20). For states where this definition does not hold, county subdivisions are created "in cooperation with state, tribal, and local officials . . . [in ways that] usually follow visible features" (U.S. Census Bureau, n.d.). That being said, by placing distilleries into their county subdivisions, we were able to empirically describe several characteristics of the spatial contexts in which the distilleries are located. In particular we computed: (1) the fraction of area in each county subdivision dedicated to corn production, (2) the fraction of area in each county subdivision characterized by limestone geology, and (3) the fraction of population in each county subdivision characterized by the U.S. Census Bureau as "rural." While the former two variables describe physical geographic characteristics that represent the primary ingredients in bourbon, the latter is intended to describe the type of human settlement in which a distillery is situated. In doing so, we note that the Census Bureau's definition of "rural" is problematic and far from definitive (Lang 1986); however, we adopt it here to paint a partial, inchoate picture of the types of settlements in which bourbon distilleries are located.

To accomplish the objectives laid out above, geologic data were obtained from the U.S. Geological Survey to identify areas that have limestone as their primary or secondary geological composition. Because limestone-filtered water is used as a selling point for Kentucky bourbon, it is worthwhile to investigate whether or not non-Kentucky distillers locate in areas with similar geology. Corn production data were obtained from the U.S. Department of Agriculture, and population data were obtained from the U.S. Census Bureau.

Results

Content Analysis

Our findings include several results related to the characteristics selected for study, namely place-names, historic names, instances of history or heritage, and references to water and local grain use. In sum, Kentucky distilleries seem to have a very strong connection to history and heritage, as more than 80 percent of those distilleries studied feature discussions of history on their websites. Compared to non-Kentucky establishments, a higher percentage of Kentucky distillers use a place-name as part of their branding, once again indicating a more pronounced connection to place than found in other distillers. A full 50 percent of Kentucky distillers have a historic reference in their name, compared to only 29 percent of non-Kentucky distillers. Again, this indicates a connection to the roots of bourbon production in Kentucky. Water was mentioned by 40 percent of Kentucky distillers and only about 19 percent of non-Kentucky distillers. Local grains, however, were identified by approximately 65 percent of non-Kentucky distillers and only 40 percent of Kentucky distillers. This may be an indication of the decline of corn production in Kentucky since the 1800s, but it could also be a way for non-Kentucky distillers to establish a connection to place by mentioning the use of local products for their bourbon.

Physical Geography and Spatial Analysis

Figure 12 maps the distribution of (1) corn production, (2) limestone geology, and (3) distilleries in the conterminous United States. It also shows the geographic mean center (GMC) of the distilleries that we studied. The GMC is the unique point created by obtaining the average

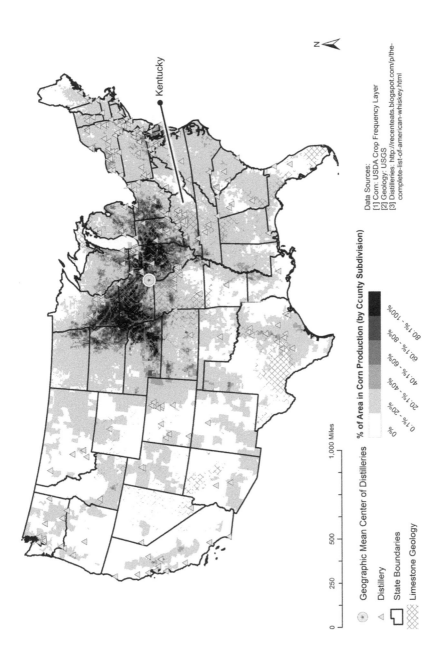

FIG. 12. Geography of bourbon distilleries in the conterminous United States. Map produced by Christopher R. Holtkamp, Brendan L. Lavy, and Russell C. Weaver using data from the USDA Crop Frequency Layer, USGS, and *Sku's Recent Eats* (blog), http://recenteats.blogspot.com/p/the-complete-list-of-american-whiskey.html.

Kentucky

N

Data Sources:
[1] Corn: USDA Crop Frequency Layer
[2] Geology: USGS
[3] Distilleries: http://recenteats.blogspot.com/p/the-complete-list-of-american-whiskey.html

% of Area in Corn Production (by County Subdivision)

80.1% – 100%
60.1% – 80%
40.1% – 60%
20.1% – 40%
0.1% – 20%
0%

⊛ Geographic Mean Center of Distilleries
△ Distillery
⬜ State Boundaries
▨ Limestone Geology

0   250   500   1,000 Miles

TABLE 5. Median values of key variables in Kentucky-based and non-Kentucky county subdivisions that contain bourbon distilleries

	KENTUCKY	EVERYWHERE ELSE (CONTERMINOUS U.S.)	p
Percentage corn area	1.4	0.0	0.014*
Percentage rural	23.0	6.0	0.022*
Percentage limestone	61.9	0.0	<0.001***

Created by authors with data from USDA Crop Frequency Layer, USGS, and *Sku's Recent Eats*, http://recenteats.blogspot.com/p/the-complete-list-of-american-whiskey.html.

Cell entries are variable medians; p-value is from Wilcoxon tests for equality of medians in independent samples

*p < 0.05 ***p < 0.001

x-coordinate and average y-coordinate of the distilleries. In other words, it gives something of the "typical" location of a distillery in the United States. Observe that while this location does not currently lie within the state of Kentucky, it exists well within the largest corn-producing region of the United States, and it is situated on a large, contiguous band of limestone geology.

The corn production layer depicted in figure 12 is a summary of the 2016 USDA "crop frequency" dataset at the level of the county subdivision. It shows the fraction of a subdivision's area dedicated to regular corn production, which is defined as an area where corn was planted in at least four years from 2008 to 2016. The decision to use four years as the baseline is intended to show regular corn production regions. By setting our threshold at four years, we capture all spaces represented in the dataset that plant corn *at least* every other year (e.g., 2009, 2011, 2013, 2015). In that vein, the locations shown in figure 12 produce corn consistently. With this point in mind, observe that many non-Kentucky distilleries are located outside of corn-producing regions. Table 5 summarizes the median differences between Kentucky-based and non-Kentucky distilleries on three variables: (1) percentage of county subdivision area dedicated to corn production, (2) percentage of county subdivision area featuring limestone geology, and (3) percentage of county subdivision population classified by the U.S. Census Bureau as "rural." As shown in table 5, nonparametric Wilcoxon/Mann-Whitney tests for equality of medians in independent samples suggest that Kentucky-based distilleries are found in places that produce significantly more corn, contain significantly more limestone geology, and are significantly more "rural" than distilleries outside of bourbon's home state.

Discussion

New Physical Landscapes of Bourbon

As seen in figure 12, despite bourbon being most closely associated with Kentucky, it has truly become America's whiskey; it is produced in nearly every state and in landscapes far removed from the limestone and Bluegrass region of Kentucky. As bourbon distillation has become geographically widespread, the association with Kentucky has diminished; however, producers are creating new identities and a new bourbon heritage based on local attributes and characteristics. For example, even outside of Kentucky some producers explicitly celebrate their location: Van Brunt Stillhouse calls out its location in the "waterfront neighborhood of Red Hook in Brooklyn" (Van Brunt Stillhouse n.d.), while others celebrate how their location influences the quality of their liquor. Boatwright Bourbon, distilled by Port Chilkoot Distillery, includes this statement in the description of its bourbon: "After distilling, we age Boatwright Bourbon for two years in barrels made of charred Kentucky White Oak. Something very Southeast Alaskan happens during that process: the swings in barometric pressure that are common here cause our barrels to slightly contract and expand. Each time that happens, the barrel absorbs and releases the slowly aging bourbon, adding nuance, depth, and flavor notes you won't find anywhere else. So when the skies turn cloudy, we smile. We figure somebody up there really loves great bourbon" (Port Chilkoot Distillery n.d.). Approximately 44 percent of distilleries include a place-name in the name of the distillery or the bourbon, indicating the importance of place; however, that place is no longer necessarily rooted in the agrarian past but instead is linked to the specific place where the distillery is located. As has been noted, "looking for the sense of and connection to place is behind the strong pull of hometown loyalty and yearning that encourages people to buy" the local product (Holtkamp et al. 2016, 66). New distilleries embrace place as an attribute in the quality of their bourbon, as described above, and they also celebrate the history of particular places, as will be discussed below.

One area where the physical landscape of production retains a connection to the origins of bourbon production is in the prevalence of local grains. Recall that one of the reasons Kentucky became synonymous with bourbon was the availability of high-quality corn (Minnick 2016). Despite the historic connection between Kentucky corn and bourbon production, only 40 percent of Kentucky distillers mention local grains, while 65 percent of non-Kentucky distillers do. This may reflect the decline in corn production in Kentucky, or possibly the scale of production for most Kentucky distillers requires a larger supply of grain than can be acquired locally. The use of local grains reinforces the distilleries' connection to place and also ties into consumer interest in local products and knowing where their food and drink is produced. Local products also allow for a more unique product, such as the Blue Corn Bourbon produced by Don Quixote Distillery in New Mexico. Blue corn is closely associated with the Southwest, which reinforces the connection to place while also creating a unique spin on a bourbon made from it.

Although bourbon enthusiasts tout the role of limestone-filtered water in producing the best bourbon, water was not mentioned in the large majority of producers' web presence. Whereas 40 percent of Kentucky distillers mention water, less than 20 percent of non-Kentucky distillers include such a mention on their websites. References to water outside of Kentucky primarily focus on the purity of the water used in distillation rather than the benefit of limestone filtration. Black Canyon Distillery in Colorado mentions pure Rocky Mountain water, while

Dancing Pines Distillery, also in Colorado, touts its snowmelt water. Others simply mention the quality of local water, as with Grand River Spirits in Carbondale, Illinois, which highlights the nationally top-ranked water available from the city water system. Although only a relatively small percentage of distillers mention water in their online promotional verbiage, those that do so recognize the importance of water in the quality of bourbon and celebrate that connection when they can.

New Cultural Landscapes of Bourbon

Heritage is a celebrated aspect of Kentucky bourbon, with more than 80 percent of Kentucky distilleries explicitly making a connection to local history and bourbon heritage in their online presence. Outside of Kentucky less than 30 percent of distilleries included any historical discussion in their marketing text. Those that did expressed a connection to local history and identity rather than bourbon heritage. An example is Quincy Street Distillery in Riverside, Illinois. This producer tells the story of Cook County sheriff Stephen Forbes creating a militia and a subsequent celebration that included punch made in a bourbon barrel and cooled in a nearby spring, hence the brand name Bourbon Spring. These kinds of references to local history reinforce connection to place and shared identity for distilleries.

A compelling pattern emerges when examining where history and heritage are included in marketing. There appears to be a predominance of historical acknowledgment in areas that have a history of liquor production. Distilleries in Virginia, Pennsylvania, New York, and Illinois have a higher likelihood of including historical references in their marketing than do producers located elsewhere outside of Kentucky. This indicates that these distillers recognize the value of local history and see an opportunity to create a connection between their business and the heritage of their community. Consumers are looking for those unique experiences and opportunities to engage with local identity, making it a savvy strategy for businesses seeking to differentiate themselves and attract customers (Holtkamp et al. 2016).

Other distilleries are creating a connection to place more directly—by engaging with and celebrating their communities. Cardinal Spirits in downtown Bloomington, Indiana, discusses a passion for American manufacturing and the opportunity to create local employment opportunities and support other local businesses by sourcing local ingredients. This connection to manufacturing is a nod to the reality that spirit production, even at a small scale, is really a manufacturing process. It also connects to the era when bourbon distillers highlighted this side of production rather than the more typical agrarian connections presented today. As one scholar notes, "Many whiskey brands up until the mid-twentieth century boasted that their distilleries were industrial factories" (Mitenbuler 2015, 254). This industrial ideal has largely disappeared. The new marketing image craft distillers embrace is that of "grain-to-glass," in which distillers celebrate the fact that they grow the grain used in their bourbon. The new image points to agriculture as the primary identifier rather than the manufacturing aspect of production.

Although the true story of bourbon may never be agreed upon, the history, heritage, and tall tales embraced by many bourbon distillers as marketing tools connect the product to the places where it is produced; in other words, old and new bourbon distillers are leveraging local and regional characteristics to create their unique identity in the marketplace. Dry Fly Distilling in Spokane, Washington, for example, is named for the owners' love of fly-fishing and is a nod to the proximity of excellent fishing. Veteran-owned distilleries play up their connection to military service as part of their identity. Most distilleries are actively engaged in their communities and participate in charitable giving programs. Dancing Pines Distillery in Colorado not only financially supports local charities but pays its employees for forty hours of volunteer work annually. These activities not only enhance connection to place but are rewarded by consumers who are choosing to support businesses that engage in these practices (Tilley, Hooper, and Walley 2003). As discussed in detail in chapter 2 of this volume, distillers are producing a public good through the production of place connection and shared identity as they build a relationship with their consumers.

In sum, although bourbon production is still dominated by a few megadistilleries operating in Kentucky, craft distillers are opening at a rapid pace. These distillers are embracing local attributes and heritage to create their identities rather than embracing the existing heritage of bourbon. They are creating new cultural landscapes of bourbon that celebrate the unique locations where it is being produced, whether it is in the heart of Kentucky Bluegrass country or on the seashore in Oregon. These far-flung producers are creating a new image for bourbon that builds on the foundation of bourbon history while creating a new identity to appeal to today's consumers.

Conclusion

As presented throughout this volume, producers of fermented products are affecting, and affected by, the physical and cultural landscapes in which they find themselves. Bourbon is no exception. After decades of bourbon being "your grandfather's drink," new producers are moving out of the shadow of the rich heritage of Kentucky bourbon to strike a new identity for themselves. This is not a rejection of their shared history; instead these producers are embracing the unique identity and attributes of their own places. This connection to place attracts consumers wanting a product unique to their experience but with a connection to the larger history of bourbon. Although bourbon is now being produced everywhere from gritty urban neighborhoods to rural idylls, it remains America's spirit, reflecting the diversity of the country while maintaining a connection to its roots, both cultural and physical.

Notes

1. Unscrupulous producers would use colorants and additives, even toxic things like turpentine, to flavor their alcohol, which they would then call "bourbon" (Mitenbuler 2015).

References

Allen, Frederick. 1998. "The American Spirit." *American Heritage* 49 (3): 82–92.

Clark, T. D. 1929. "The Ante-Bellum Hemp Trade of Kentucky with the Cotton Belt." *Register of the Kentucky State Historical Society* 27 (30): 538–44.

Fryar, Alan E. 2009. "Limestone Water and the Origin of Bourbon." Presentation at the Geological Society of America annual meeting, Portland OR, October 18–20, 2009, *GSA Abstracts with Programs* 41 (7).

Holtkamp, Christopher, Thomas Shelton, Graham Daly, Colleen Hiner, and Ronald Hagelman. 2016. "Assessing Neolocalism in Microbreweries." *Papers in Applied Geography* 2 (1): 66–78.

Hsieh, Hsiu-Fang, and Sarah Shannon. 2005. "Three Approaches to Qualitative Content Analysis." *Qualitative Health Research* 15 (9): 1277–88.

Kemmerer, Donald. 1949. "The Pre–Civil War South's Leading Crop, Corn." *Agricultural History* 23 (4): 236–39.

Kentucky Distillers' Association. 2016. "Kentucky Bourbon Trail® Visits Skyrocket With 900,000 Guests in 2015." January 21, 2016. https://kybourbon.com/kentucky-bourbon-trail-visits-skyrocket-with-900000-guests-in-2015/.

Krippendorff, Klaus. 2013. *Content Analysis: An Introduction to Its Methodology*. 3rd ed. Los Angeles: SAGE.

Kroll, Harry H. 1967. *Bluegrass, Belles, and Bourbon: A Pictorial History of Whisky in Kentucky*. South Brunswick NJ: A. S. Barnes.

Lang, Marvel. 1986. "Redefining Urban and Rural for the U.S. Census of Population: Assessing the Need and Alternative Approaches." *Urban Geography* 7 (2): 118–34.

McClay, Wilfred, and Ted McAllister. *Why Place Matters: Geography, Identity, and Civic Life in Modern America*. New York: Encounter Books.

McKeithan, Seán. 2012. "Every Ounce a Man's Whiskey? Bourbon in the White Masculine South." *Southern Cultures* 18 (1). http://www.southerncultures.org/article/every-ounce-a-mans-whiskey-bourbon-in-the-white-masculine-south/.

Minnick, Fred. 2016. *Bourbon: The Rise, Fall, and Rebirth of an American Whiskey*. Minneapolis MN: Voyageur Press.

Mitenbuler, Reid. 2015. *Bourbon Empire: The Past and Future of America's Whiskey*. New York: Penguin Books.

New York Times Magazine. 2003. "The Way We Live Now: The (Agri)Cultural Contradictions of Obesity." October 12, 2003.

Port Chilkoot Distillery. n.d. "Whiskies and Spirits." Accessed May 5, 2018. http://www.portchilkootdistillery.com/whiskies-spirits/#whiskies-spirits-1.

Simpson, Lee. 2008. "Louisville—A City with History." *Public Historian* 30 (4): 88–91.

Tilley, Fiona, Paul Hooper, and Liz Walley. 2013. "Sustainability and Competitiveness: Are There Mutual Advantages for SMES?" In *Competitive Advantage in SMES: Organising for Innovation and Change*, edited by Oswald Jones and Fiona Tilley, 71–84. London: John Wiley and Sons.

U.S. Census Bureau. n.d. "Geographic Terms and Concepts—County Subdivision." Accessed May 5, 2018. https://www.census.gov/geo/reference/gtc/gtc_cousub.html.

Van Brunt Stillhouse. n.d. "About." Accessed May 5, 2018. http://www.vanbrunt stillhouse.com/.

Weaver, Russell, and Christopher Holtkamp. 2016. "Determinants of Appalachian Identity: Using Vernacular Traces to Study Cultural Geographies of an American Region." *Annals of the American Association of Geographers* 106 (1).

Whiskey Advocate. 2017. "A Timeline of Bourbon History." December 14, 2017. http://whiskyadvocate.com/bourbon-timeline/.

Apples and Actor-Networks   5

Exploring Apples as Actors in English Cider

Walter W. Furness and Colleen C. Myles

> It may be questioned whether our peasantry, in the aggregate, could now toil through the heats of summer without the aid of fermented liquours.
>
> —Thomas Andrew Knight, *Pomona Herefordiensis*

Cider Renaissance

Through the first half of the 2010s cider was the fastest-growing sector of the alcoholic beverage industry in the United States, and it also experienced significant growth globally (AICV 2017; NACM 2016; Petrillo 2016). While this industry umbrella includes other products, such as nonalcoholic ciders and cider vinegar, the vast majority of growth has stemmed from the increasing popularity of hard cider, which has risen alongside the surging craft beer movement (Petrillo 2016). In response to this trend, both small-scale producers and global beverage corporations have moved to expand their offerings of craft ciders, though debates over authenticity and definitions figure prominently in fermented beverage discourses.[1] Around the world large producers such as Heineken, Anheuser-Busch InBev, and MillerCoors now dominate the cider industry, having quickly leveraged economies of scale and name recognition to produce new "craft" products and acquire existing ones (Petrillo 2016). Large beverage corporations in Europe and elsewhere are also capitalizing on this growth by bringing more cider products into their respective folds (Treaner 2003; Scott 2008; Gerrard 2018). Nevertheless, the notion of "craft" remains important, even in places like Herefordshire, which is home to cider producers of all scales. While the cider industry's once-breakneck growth has plateaued and is expected to experience only modest gains in the immediate future, this recent renaissance has transformed its profile indelibly (NACM 2016; Petrillo 2016).

In this chapter we use the term "cider" to refer to alcoholic or "hard" cider produced from apples and, to a lesser extent, pears (a product known in England as perry). Although traditional cider production historically has been most closely associated with England's Three Counties region (Herefordshire, Gloucestershire, and Worcestershire) (fig. 13), other strong regional traditions thrive despite the limited attention they receive outside of Europe, particularly from Americans. Some examples include *cidre* in the Bretagne and Normandie regions of France (Jenkins 2016), *apfelwein* in Hesse, Germany (AICV 2017), and *sidra* in Asturias, Spain (Dapena, Miñarro, and Blázquez 2005). Building upon long-standing cultural traditions, cider's recent surge in popularity has raised its profile and expanded production from its established hearth(s) in Europe to the broader world.

Contextualizing English Apple Cider vis-à-vis Fermented Landscapes

Cider is particularly placed in the world of fermented beverages: although it is more often seen served alongside beer than wine, numerous aspects of its production tend to more closely mirror viticulture. For instance, while the malted grains and hops used in beer production may be grown "from scratch" in a single season, orchards for cider apples or pears can take years to establish and maintain (Knight 1811). Moreover, as one English grower mentioned during an interview in 2016, "cider apples are hard to do anything with other than make cider . . . so orchards are planted under contract in direct relationship with cider makers," thus guaranteeing the sale of a given harvest and assuring the salability of a crop that is not otherwise marketable. Indeed apple producers commonly contract with multiple cider makers, given the inten-

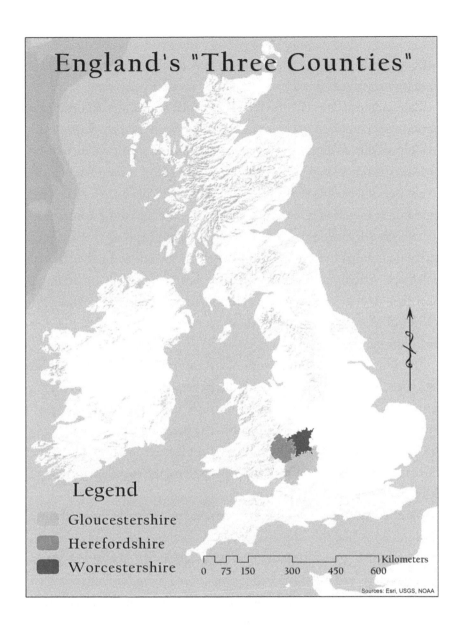

FIG. 13. The location of England's "Three Counties" (Herefordshire, Worcestershire, Gloucestershire). Map created by Walter W. Furness; some basemap data are from Esri, USGS, and NOAA.

sive resource input required to establish and maintain orchards. Even with these efforts, in many cases producers do not break even on their orchard investment until seven to nine years after planting.

This embeddedness in the landscape adds legitimacy to claims of *terroir* often employed by cider makers, who nevertheless employ numerous approaches to crafting a "local" product. Like wines, ciders may be produced from a single varietal of fruit or a blend incorporating different desirable characteristics (NACM 2016; Petrillo 2016). Although numerous varietals of bittersweet apples still exist (i.e., those apples used specifically for cider as opposed to those meant expressly for eating), today their cultivation is largely a matter of concern to small artisanal producers striving to highlight a particular flavor profile or heritage. This claim is evidenced by the fact that many producers can name only a handful of varietals offhand and can recognize only a modest number of additional types (Reedy et al. 2009).

The United Kingdom's National Association of Cider Makers (NACM) defines cider as an alcoholic beverage, explicitly referencing fermentation as a key process to be managed in the production of quality cider: "Cider is an alcoholic drink: The product of fermented apple juice. With hundreds of different cider apple varieties available to the cider maker there is an almost limitless potential to produce drinks of different styles" (NACM n.d.).

Founded in 1920, the NACM also serves as an umbrella for local cider organizations like the Three Counties Cider and Perry Association, the South of England Cidermakers Association, the Welsh Perry and Cider Society, and Westons Cider, the United Kingdom's fifth-largest producer. Alongside cider, the association also promotes perry, touting its "illustrious history" and noting its cultural hearth in the counties of Gloucestershire, Herefordshire, Worcestershire, and Somerset while distinguishing traditional perry—produced from specific varietals—from "pear cider," made with nontraditional varietals (NACM n.d.). The NACM makes explicit the connection between culture and landscape via the material medium of the apple, claiming that "cidermakers' heritage, community focus, innovation and support for the economy, both rurally and nationally, make cider an integral part of our culture and landscape" (NACM 2017).

Given the relative lack of literature dealing explicitly with cider compared to its more prominent alcoholic cousins (e.g., wine and beer), there is space to more thoroughly explore cider as a distinct beverage with its own rich geographies, cultures, and economies. In England in particular the topic is ripe for the picking (pun intended) using a lens of fermented landscapes because, even while British cider constitutes a near majority of the global market for cider (45 percent), domestic apple production lags behind cider production, as evidenced by the fact that only 56 percent of the apples used in this British cider are grown in-country (NACM 2017). This gap between production and sourcing has implications for how we conceptualize what is a local and culturally significant commodity.

As outlined in the opening chapter of this book, the "fermented landscapes" research program is one that focuses on the production, distribution, and consumption of fermented foods and beverages in addressing rural-urban exchanges or metabolisms over time and space as well as ongoing processes of cultural and environmental change and material transformation. Thus we ask this question: What can a glimpse into the English apple cider industry—from branch to bottle—tell us about the extant relationships between scale, power, and difference in a local-global product?

Data and Methods

This chapter draws upon fieldwork conducted in Herefordshire, England, in early May 2016, including walking interviews ("walkabouts"), more traditional individual and group interviews, and photographs. We have mixed these primary sources with existing literature and analysis of industry reports as well as promotional content from cider producers themselves. The producers selected are all headquartered in Herefordshire and include a small-scale cider producer who distributes only locally (Gregg's Pit), a medium-scale producer who distributes both regionally and internationally (Westons), and a global-scale producer that crafts commercial-industrial cider (Bulmers).

We inquire in this particular context about the linkages between the material and the semiotic in English apple cider production to explore the multiscalar meanings of and implications for cider production in this globalizing industry. In order to try to understand the myriad actors and implications of the complex local-to-global production and distribution patterns within English cider, we apply an actor-network theory (ANT) approach to investigate how the actor-networks between different models and scales of production, distribution, and consumption differ, with the ultimate goal being explanation: What can be said about this collective? Specifically, how does scale bear on the relationships between producers and product? In what ways are the apples acting in these assemblies?

While actor-network theory has become a "disparate family of material-semiotic tools, sensibilities, and methods of analysis" in recent years (Law 2007), in this chapter we use an ANT approach to explore what insights may be gained through a careful application of its perspective to Herefordshire cider. The use of an empirical case study is necessary to do justice to the true sense of ANT, which is not a foundational theory but rather a descriptive tool for understanding how relationships of things assemble or do not assemble (Latour 2004; Law 2007). From a Latourian perspective, ANT offers a framework through which we, as followers of the apple, may locate the most important articulations of power and meaning ("matters of concern," to borrow a term from Bruno Latour) within the complex realities of multiscalar production, consumption, distribution, and definition. Far from a well-mapped, discrete network in the technical sense, the network (or more accurately, *agencement*/assemblages) we seek to elucidate here is not so easy to pin down (Latour 1996).

Findings and Discussion: On the Possibility and Peril of "Two Different Worlds"

The cider production industry encompasses enterprises related to growing fruits (primarily apples and pears but also, less commonly, various nontraditional fruits) and transforming them into alcoholic and nonalcoholic cider beverages (Petrillo 2016). Herefordshire's long history of cider fruit production is well documented, particularly by the *Pomona Herefordiensis* (Knight 1811), a monumental work that includes detailed color engravings of thirty varieties of apples and pears used for making cider and perry (fig. 14). Thomas Andrew Knight's volume is one of a number of catalogs of cider fruit varietals and features not only the detailed illustrations of each variety but also advice on breeding new varieties, orchard management, and processing the fruits into cider and perry.

It is worth noting that the Herefordshire cider industry is geographically indicated. As one grower told us in 2016, "All of our products have protected geographical indications [PGIs].... All of them named varieties, all of them have the vintages on, and you can't use the term 'Herefordshire cider' or 'Herefordshire perry' unless you have a PGI." In addition, and importantly, cider production in Herefordshire and elsewhere in the United Kingdom is shaped in no small way by economic as well as geographic factors. Specifically, the UK tax code creates the conditions for small cider makers (in this case, those producing fewer than seventy hectoliters [hl] per year) to operate in a relatively free manner, unperturbed by more intricate regulatory requirements exacted on larger-scale producers (Gov.uk 2018). This provision also creates a sort of natural barrier to expansion for smaller producers looking to branch out.

As a result, the progression from "small" to "big" in English cider amounts to a leap rather than a steady climb due to taxation policies. Figure 15 shows the large number of producers just below the seventy-hectoliter limit, with the next sizable number coming at the thousand-hectoliter level. This striking visual is a product of economies of scale and a tax code that allows cider makers producing fewer than seventy hectoliters in a rolling twelve-month period to claim exemption from

The Redstreak L.

FIG. 14. Illustration and text from Thomas Andrew Knight's (1811) *Pomona Here-fordiensis*. Natural History Museum Library, London.

registering with Her Majesty's Revenue and Customs (HMRC) for duty purposes (Gov.uk 2018). Given the extra effort and financial resources necessary to comply with registration and duty requirements, smaller producers strive to increase production rapidly once they exceed the seventy-hectoliter threshold in order to continue to thrive at higher

POMONA HEREFORDIENSIS.

I.

THE REDSTREAK.

> ———" *whose pulpous fruit*
> " *With gold irradiate and vermilion shines.*" PHILIPS.

The Redstreak, or Redstrake as our ancestors wrote it, appears to have been the first fine cider apple that was cultivated in Herefordshire, or probably in England ; and it may even be questioned whether excellent cider was ever made, in any country, previous to the existence of this apple. The Redstreak is unquestionably a native of Herefordshire, and the credit of having raised it from seed, in the beginning of the 17th century, has been generally given to Lord SCUDAMORE, from whom it was, when first cultivated, called " *Scudamore's Crab :*" and if that nobleman did not raise it from seed, it appears extremely probable that he found the original tree growing upon his estate ; and he certainly first pointed out the excellence of the fruit to the Herefordshire Planters.*

Trees of the Redstreak can now no longer be propagated ; and the fruit, like the trees, is affected by the debilitated old age of the variety, and has in a very considerable degree

* Evelyn's Pomona.

---

taxation levels (Gov.uk 2018). This also creates a bifurcated distribution of cider production, as smaller companies muster their resources to make the jump from small-scale operations to the threshold of larger-scale productivity and the relevant rules and tax code requirements. These regulations include four different duty categories for cider based on its

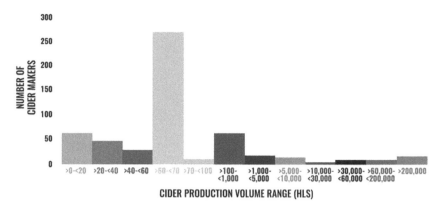

FIG. 15. UK cider production in hectoliters by number of producers. National Association of Cider Makers, http://cideruk.com/uk-cider-market/. Used with permission.

percentage of alcohol by volume (ABV) and whether it is sparkling or not, as well as reporting requirements (Gov.uk 2018).

Through our fieldwork, we identified a common concern among smaller producers: the fear that this special seventy-hectoliter tax shield will be removed or modified in the future. One study participant noted in 2016 that, if smaller producers had to conform to the taxation guidelines of larger producers, "all of those [smaller] cider makers would stop making cider . . . because it wouldn't be worth . . . registering with the tax authorities. . . . Everything would have to be approached in a different way. It's just not worth the effort." In fact some producers are nervous about how a potential removal of tax protections would impact cider makers overall, pointing to the broader benefits to the industry associated with well-known brands. One producer noted, "If you remove that layer, all the heritage and traditional values that consumers understand from the cider industry, that they get from the craft industry, go out the window. So if you make Strongbow [a cider, part of the Bulmer group now owned by Heineken], you don't need to tell the consumer that there's a huge craft and heritage cider tradition in the UK because . . . they've seen it. . . . [In] fifty years' time, if you've taken out the seventy-hectoliter [the smaller producers' typical limit], Strongbow [brand cider] is just another drink."

Such effects are of course economic but also manifest in the landscape. While traditional, "standard" English orchards are multiuse spaces with high canopies to allow cattle and sheep to graze underneath, more modern or "bush" orchards have lower canopies and more uniform designs to facilitate higher yields and the robotic flails of mechanized harvesters. While traditional orchards are important spaces of biodiversity and historic rootstocks, they are also costly to prune and maintain (Johnson 2010). Consequently, the extent of traditional orchards in the United Kingdom decreased over the second half of the twentieth century, despite increased interest in their preservation by some craft cider makers (Johnson 2010). Smaller producers tend to worry that if their industry segment disappears, so too will the iconic orchard landscapes that help make the rural English countryside what it is: "If they stopped making cider, why would they leave those standard orchards there?" one interviewee told us. "So you'd lose a huge chunk of the landscape because all of a sudden the government has done something dumb. . . . And that would be such a shame." Producers must contend with the "wildness" of apples in traditional orchards, balancing productivity, efficiency, heritage, and quality.

With the current policy, stated one interviewee, "it's almost like there's two different worlds": those who produce under the taxable limit and those producing above the limit. While some producers (such as Westons) have managed the jump from smaller-scale to larger-scale production while maintaining their local and artisanal identity, this often proves a difficult route to travel, as our interviewee told us: "That's what's really interesting about Westons . . . [it] was tiny. If you wanted to taste cider, someone would shuffle out from the back [for the tasting] . . . it was just a small farm operation before it became big." Predictably, many smaller producers see the shift to "big" as problematic in terms of quality maintenance—both for consumer experiences and for the product itself.

This difference is frequently seen as one of quality and honesty. This is how one small-scale producer described it in our 2016 interview:

PRODUCER: [explaining his methods]: So all you need is sugar from the sun, wild yeast, [and] bingo. You do not need tanker loads of corn syrup. You do not need malic, ascorbic, brown coloring, various other things that don't go on the label, etc., etc. And you do not need to ferment at the speed of a high-speed train for three weeks in two-hundred-thousand-liter vats. . . . [Here] the sugar is coming from the sun, and we're . . . using long fermentations, typically nine months, sometimes longer. . . . So [our process is] different. Not better, but different.

INTERVIEWER: [countering with an experience at a nearby large-scale producer]: "Not better, not worse, just different," is the exact opposite of what our tour guide [at Westons] said. . . . "Some people are right and some people are wrong, and we're right"—that's exactly what she said.

PRODUCER: Well, [their process] is not honest. When I talk about difference, what I'm really talking about is honesty. What you see on that label is what you're getting.

In contrast to how this small producer describes Westons (the United Kingdom's fifth-largest cider producer, now a much more commercialized operation), his own tour/tasting is very "homegrown," pun intended (fig. 16).

The notion, often repeated in field interviews, of "not better, not worse, but different" exemplifies the notion of generalized symmetry in Michel Callon's (1984) "sociology of translation," referring to the commitment to explaining conflicting viewpoints in the same terms. In other words, each producer has a preferred modus operandi, but each is competing in the same market, and, moreover, their individual paths are interrelated. As one producer put it in 2016, "There's room in the market for everyone. The big guys did us all a favor by growing the way they did during the Magners and Bulmers [both large-scale, international producers] heyday," because it improved the overall market for English cider. "And now we're in a situation where the craft boys are doing stuff, interesting things, that the big boys don't understand" but that they nevertheless want to emulate.

FIG. 16. Orchard tour and tasting at a small-scale apple cider producer in Hereford-shire, England. Photo by Colleen C. Myles, May 7, 2016.

This is not unlike other emerging—though, ironically, historic—fermented landscapes. In the Sierra Nevada foothills of California, for example, where wine grape growing and winemaking are an increasingly prominent part of the socioeconomic and physical environmental context, the presence of "big guys" has proven useful for smaller producers—just as the presence of the small producer has helped the big ones—via a process and commitment, whether implicit or explicit, to *coopetition* (Myles and Filan 2019). As such, in place of imposing a priori assumptions about actors, scales, and labels (e.g., global versus local), one can take a different approach to understanding the multiple constituents that are enrolled and held together in the world of Herefordshire cider. As suggested by Latour (1996, 371–72), "a network notion is ideally suited to follow the change of scales since it does not require the analyst to partition her world with any [a] priori scale. The scale, that is, the type, number and topography of connections, is left to the actors themselves. . . . Instead of having to choose between the local and the global view, the notion of network allows us to think of a global entity—a highly connected one—which remains nevertheless continuously local."

Not Better, Not Worse, but Different

So what can be said about the apples themselves? In the brief cases presented here, we have begun to sketch out the contours of apples as actors (actants) that "modify other actors through a series of trials that can be listed thanks to some experimental protocol," to use a basic Latourian notion (Latour 2004, 75). Their behavior is evident through their performances (evinced through the aforementioned trials of interviews and observation), which point toward their competences (their properties as actors) (264).

One behavior we have noted is the bearing the apples have on matters of definition. For example, in the case of French *cidre*, the phrase "100% pur jus" is an important label marking. According to a small-scale English producer, using 100 percent "named varieties" both creates a definitionally pure cider and imbues it with value. While this case argues that valorization produces difference over dominance ("not better, not worse, but different"), it nevertheless showcases one way in which the apples act in this assembly. This perspective reimagines traditional notions of scale and power, showing how the idea that "the craft boys are doing stuff . . . that the big boys don't understand" reflects more nuanced relationships and mutualism than Darwinian competition.

From this perspective numerous possible lines of inquiry begin to materialize. How is this matter of modifying other actors through apples' definitional influence shared or contested by other producers of varying sizes? Do the methods used by larger producers to extend and maintain networks of cider across different spaces and territories co-opt or push out smaller producers? Or do they rather pull them along because, as one producer noted, "there's room in the market for everyone, and the big guys did us all a favor by growing the way they did"? Does the apples' influence shape power relations between producers by valorizing products differently, allowing for niches of difference instead of the conventional wisdom of dominance? It is evident that the apples already shape the discourses that producers engage in, but this could also be the subject of further examination given its bearing on competing notions of the local and global and what those labels ultimately mean for cider makers in England and elsewhere.

Clearly, the points of departure for tracing various matters of concern are numerous. So what can ANT offer in the way of signposts through this imbroglio? First, it offers us the key ground-level perspective that has helped materialize many of these questions in the first place (via identifying matters of concern). This approach differs from a political economy perspective that brings more a priori assumptions to the discussion regarding producer size and power dynamics, and it also provides the benefit of subverting the subject-object impasse that often weighs on political ecology (see Latour 2004). Yet with this blessing comes a danger to the disciple of ANT: that of infinite derailment in the face of multiplying matters of concern. Thus, care must be taken to maintain a focused and disciplined approach to following the questions one wishes to examine. This has been evident in our own work on this chapter, as ever more possible trajectories of inquiry open up along the way.

Notes

1. See, for example, the debates over "real ale"—including cider and perry—as promoted by the Campaign for Real Ale (http://www.camra.org.uk/about-cider -perry). CAMRA's views about these traditional products conflict with typical legal and market-based definitions of what constitutes cider.

References

AICV. 2017. "European Cider Trends 2017." European Cider & Fruit Wine Association (AICV). http://aicv.org/file.handler?f=CiderTrends2017.pdf.

Callon, Michel. 1984. "Some Elements of a Sociology of Translation: Domestication of the Scallops and the Fishermen of St Brieuc Bay." *Sociological Review* 32 (1): 196–223.

Dapena, Enrique, Marcos Miñarro, and María Dolores Blázquez. 2005. "Organic Cider-Apple Production in Asturias (NW Spain)." *IOBC WPRS Bulletin* 28 (7): 161–65.

Gerrard, Bradley. 2018. "Aspall Cider Snapped Up by Molson Coors After Eight Generations of Family Business." *The Telegraph*, January 7, 2018. https://www.telegraph.co.uk/business/2018/01/07/aspall-cider-snapped-molson-coors-eight-generations-family-business/.

Gov.uk. 2018. "Excise Notice 162: Cider Production." Updated February 28, 2018. https://www.gov.uk/government/publications/excise-notice-162-cider-production/excise notice-162-cider-production#para-5-12.

Jenkins, Ashley B. 2016. "An Integrated Rural Tourism Approach to Normandy's Cider Trail." Unpublished master's thesis, Texas State University, San Marcos.

Johnson, Henry. 2010. "The Traditional British Orchard: A Precious and Fragile Resource." *Historic Gardens: The Building Conservation Directory*. Cathedral Communications. Accessed August 14, 2018. http://www.buildingconservation.com/articles/traditional-orchards/traditional-orchards.htm.

Knight, Thomas Andrew. 1811. *Pomona Herefordiensis: Containing coloured engravings of the old cider and perry fruits of Herefordshire; Accompanied with a descriptive account of each variety*. London: Agricultural Society of Herefordshire.

Latour, Bruno. 1996. "On Actor-Network Theory: A Few Clarifications Plus More Than a Few Complications." *Soziale Welt* 47 (4): 369–81.

———. 2004. *Politics of Nature: How to Bring the Sciences into Democracy*. Translated by Catherine Porter. Cambridge MA: Harvard University Press.

Law, John. 2007. "Actor Network Theory and Material Semiotics." In *The New Blackwell Companion to Social Theory*, edited by Bryan S. Turner, 141–58. Malden MA: Blackwell.

Myles, Colleen C., and Trina Filan. 2019. "Making (a) Place: Wine, Society, and Environment in California's Sierra Nevada Foothills." *Regional Studies, Regional Science* 6 (1): 157–67.

NACM. n.d. "What Is [*sic*] Cider and Perry?" National Association of Cider Makers (NACM). Accessed May 29, 2018. https://cideruk.com/what-is-cider-and-perry/.

———. 2016. "UK Cider Market." National Association of Cider Makers (NACM). Accessed May 29, 2018. http://cideruk.com/uk-cider-market/.

———. 2017. "A Great British Future for Cider." Promotional report, National Association of Cider Makers (NACM). https://cideruk.com/wp-content/uploads/2017/05/A5CiderManifesto.pdf.

Petrillo, Nick. 2016. *Bitter but Sweet: Sales of Hard Cider Have Plateaued Following Years of Unprecedented Growth*. Report provided to Texas State University by IBISWorld, May 9, 2017, in accordance with its license agreement with IBISWorld.

Reedy, David, et al. 2009. "A Mouthful of Diversity: Knowledge of Cider Apple Cultivars in the United Kingdom and Northwest United States." *Economic Botany* 63 (1): 2–15.

Scott, Mark. 2008. "Heineken, Carlsberg Nab Scottish Brewer." *Bloomberg*, January 25, 2008. https://www.bloomberg.com/news/articles/2008-01-25/heineken-carlsberg -nab-scottish-brewerbusinessweek-business-news-stock-market-and-financial-advice.

Treanor, Jill. 2003. "£278m Bulmers Buy Completes S&N U-turn." *The Guardian*, April 29, 2003. https://www.theguardian.com/business/2003/apr/29/9.

Migration and the Evolving Landscape of U.S. Beer Geographies

6

Mark W. Patterson, Nancy Hoalst-Pullen, and Sam Batzli

Untold migrations
ferment American lands;
Such legacies brew.

Beer is the third most widely consumed beverage in the world after water and tea (Patterson and Hoalst-Pullen 2014). With breweries and brewpubs opening up daily across the United States and beyond, craft beer seems to have an unbridled ubiquity that was nearly nonexistent in 2010. The Brewers Association (2017) estimates that in the United States more than three-quarters of the population lives within ten miles of a brewery, with more breweries open than at any previous time in the country's history. But the modern-day beer landscape—or "beerscape"—in the United States is not a phenomenon without historical background or precedent; it is an artifact of the persistent beer culture in America that has been evolving since before the arrival of Europeans in the late fifteenth century.

Prior to the European settlements (and colonies) that would become the United States, indigenous groups fermented grains, fruits, and other plants into various alcoholic beverages, although the amount produced and the general use(s) varied widely. By and large, alcohol production was predominantly situated in the Southwest.[1] It was largely of a form most aligned with today's wine and beer (La Barre 1938). For example, the Apaches made *tizwin*, a fermented corn beer that was generally low in alcohol content, as well as *tulipi*, an agave beer made from fermented sap that looks milky, feels viscous, and tastes like soured yeast. Generally speaking, these and other beer-like beverages were not consumed on a daily basis but reserved for special rituals and ceremonies, such as births, marriages, feasts, and deaths (Bruman 2000).

Indeed the American beerscape has evolved significantly since the time before European contact. This chapter presents a geographic and historical narrative of the American fermented landscape by looking at four distinct migrations to the North American continent and the United States.[2] Specifically, we focus on the English migration in colonial America (1600–1776), the Irish migration before and after the great potato blight (1820–60), the German migration (1840–1910), and most recently the Mexican migration (1986–2015). Whether it be the beer styles, drinking establishments, brewing legacies, or holidays linked to imbibing, each migration presents different geographic and cultural impacts on America's ever-evolving fermented landscapes.

The English Migration: Establishing American Brewing (1600–1776)

The first Europeans to migrate en masse to America were the British. For these early settlers beer was of vital importance and played an integral role in the culture. It was sustenance. It was usually safer to drink than water. And it was commonly in short supply (Phillips 2014).

When the Jamestown colony was established in 1607, one of settlers' first tasks was to brew beer. Unfortunately, the settlers found they could not make satisfactory beer with their provisions, and they sent word back to England in the form of an advertisement, seeking a brewer to join their settlement. The barley they planted in 1608 was harvested, and in 1609 their brewer, newly arrived from England, produced the first successful batch of beer. The history of Jamestown underscored the difficulties early English settlers faced regarding the production of beer, namely the lack of ingredients (Keppel 2018).

With trade increasing between the New World and the Old World, beer and its raw ingredients maintained their status as treasured cargo on transatlantic voyages. As the brewing process and addition of hops inhibited the growth of various microbes (Vriesekoop et al. 2012), beer remained an onboard provision, even if it sometimes spoiled en route. Beer transported from Britain often occupied cargo space on ships, which usually carried enough beer for the crew to drink on the return voyage as well. When the *Mayflower* was blown off course from its original destination of Virginia, for example, the captain had the passengers disembark at Plymouth Rock in the New World so that the crew would have enough beer to make the voyage back to England (Chapelle 2005).

Given that beer was vital for settlers' survival, there was a huge incentive to be able to produce beer in sufficient quantities—especially from ingredients grown in the colonies. While most households produced their own beer, breweries were fundamental to settlements. The Dutch beat the British in establishing the first brewery in the new colonial area, in 1612–13. Hans Christiansen and Adrian Block opened their brewery in New Amsterdam, now part of the southern portion of modern-day Manhattan in New York City (G. Smith 1998). The location of the brewery took full advantage of its access to the North River (now called the Hudson River) for transporting ingredients and the finished product. For the Dutch, however, their control over New Amsterdam did not last; the Dutch governor surrendered the settlement to England in 1664.[3]

The 1630s witnessed large numbers of Puritans migrating to the American colony, with the majority of them settling in New England (Brooks 2017). Boston and Salem were the top destinations. While the Puritans did not condone public drinking, they did recognize the important role beer played in survival, particularly when trying to establish communities (Phillips 2014). Two of the first tasks for early Puritan settlers were building a place of worship and brewing beer. Beer, and alcohol in general, was described as a "good creature of God," although a drunkard was "of the devil" (O'Brien 2015).

Farther south, in the colony of Georgia, Gov. James Oglethorpe realized that developing this colony would require hardworking people and that alcohol could lead to low productivity. In 1735, just two years after the colony was established, Oglethorpe banned spirits (Kobler 1973) but permitted beer, which he viewed as a healthy drink. To attract settlers to his colony, he offered each settler forty-four gallons of beer (Stinehour 2016). In 1738 the first brewery in the Deep South was opened, on Jekyll Island, to help provide settlers with their beer rations. But along with forty-four gallons of beer, Oglethorpe also included sixty-five gallons of molasses, which settlers quickly fermented and distilled into a rum (Lundmark 2015). The illegal moonshine was the alcohol of choice for Georgia, in contrast to beer's popularity in the other twelve colonies.

As the population in the thirteen British North American colonies grew, the production of beer by local breweries increased. Figure 17 shows the location of breweries in colonial America from 1612 to 1776 (Batzli 2014). The majority of the 197 breweries were found along the coast, with higher concentrations most evident in the larger cities of Philadelphia (69) and New York (58). Overall, Pennsylvania had the most breweries (88), followed by New York, with 76. The lone brewery in Illinois, established in the French settlement of Kaskaskia in 1765, was the first brewery outside of the thirteen colonies.

As breweries grew in size and number, another fixture emerged on the landscape—the tavern. Most if not all of the functions provided by taverns in England were transplanted to the colonies. Taverns in early colonial America (much like those in England) provided shelter for weary travelers and hosted important social functions. It was here that people gathered to share news and gossip, discuss politics, and even plan rebellions. Several taverns served as places to hold local court, especially in more rural areas. As the immigrant population increased, so too did the number of taverns, and breweries grew along with them to meet ever-growing demand.

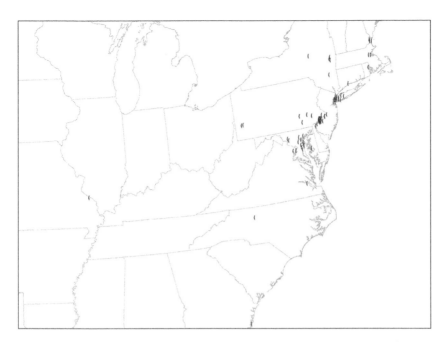

FIG. 17. Locations of U.S. breweries opened, 1600–1776. Created by Sam Batzli, 2014.

Estimates for the number of immigrants prior to the first census in 1790 was nearly a million, with approximately 425,000 of them hailing from the United Kingdom (D. Smith 1972). During this time beer in England was dominated by two styles—porter and English pale ale—which migrants were no doubt accustomed to drinking (Dighe 2016). These beers were made from barley, with the porter's malt roasted longer than that of the pale ale to impart a darker color. Barley and hops, both native to England, were essential ingredients to both ale styles. When the English migrants first started arriving in the New World, the only barley they had was what had been brought onboard ship.

Early attempts at growing barley in the New World were largely unsuccessful. In lieu of barley, early settlers often used corn, a relatively abundant native plant. Several plants were used to bitter the beer; these included wild hops, gentian root, horseradish, juniper berries, and spruce tips, among others. Hops that were cultivated in the northern to middle colonies were soon exported to other colonies, whose climate and/or soil

Migration and the Evolving Landscape   131

was not conducive for hop production. By the late 1700s New England was the largest hop-growing region along the eastern seaboard (Heisel 2011), owing largely to its more favorable climatic regime. English crop failures in the 1820s combined with improved east-west transportation networks (e.g., Erie Canal, Great Western Turnpikes) in the United States to promote hop cultivation. But by the latter part of the nineteenth century, insects, disease, mold, mildew, and drought ravaged the New England hops (Krakowski 2014). The higher-yield crops that originated in the Pacific Northwest brought competition to an already faltering market. The final blow was the growing temperance movement (and eventually Prohibition), which culminated in the collapse of New England hops production (Krakowski 2014).

Indeed the English migration laid the foundation for the beer landscape of the United States in terms of social, economic, and agrarian developments. Beer was sustenance, and its production was vital to everyday life. As the number of immigrants grew, so too did the amount of land needed to grow the crops, both native and foreign to the New World. Larger settlements promoted an increase in the number of breweries to produce beer, a task once restricted mostly to individual households. To distribute beer, taverns sprang up, and in short order these places began to help to facilitate the movement of people and provide a place for the exchange of news, information, and opinions. Although the English settlers shaped the beer landscape of their new home, the next group of immigrants had an even larger impact.[4]

The German Migration: Diaspora of the Lager (1840–1910)

If there is one group whose migration had (and arguably still has) the largest impact on the U.S. beerscape, it is the Germans. While Germans immigrated to the thirteen British North American colonies as far back as the early 1600s, their peak immigration occurred from the 1840s to 1910s. In the 1880s more than 1.4 million Germans came to the United States (DHS 2008), a time when the German economy was transitioning from agrarian to industrial. Making a living as a farmer during the rise of the Industrial Revolution became increasingly difficult, so the seemingly endless supply of land in America was a siren call to many German farmers, who brought to the United States their agricultural skills and know-how.

In order to more fully understand the impact of German migration, one must look to the early 1500s. After Columbus landed on Hispaniola (in what is now the Dominican Republic) in 1492, European monarchies and businesses were keenly interested in increasing their wealth from the resources that these New World lands had to offer. Ships would set sail for the Americas, and the voyagers focused on exploration and colonization. These ships would carry supplies—with beer being of high importance—to settlers, and they would return to Europe with New World goods.

Ships were infamous for holding stowaways, particularly microbes unbeknown to the sailors aboard. In at least one case a stowaway was the microscopic organism *Saccharomyces eubayanus*, a yeast found on the orange-colored gall fruit from the southern beech trees (*Northofagus*) of Patagonia (Libkind et al. 2011). After arriving in Europe, the New World and European yeast species used in the fermentation of ales became hybridized, creating a new yeast species that produced a new type of beer—lager (Devitt 2016). Of course at the time people knew nothing of yeast and the fermentation process, but Bavarian monks would lager this new beer in cool caves, allowing it to clarify and mellow.[5] While initially these lagers were dark brown in color (for example, bocks) and more carbonated than ales, a Bavarian brewer named Josef Groll created a light-colored lager in Bohemia (in the modern-day Czech Republic) in 1842. The lager, named after the town of Plzeň, where Groll developed it, would be called pilsner (meaning "from Plzeň"), and the style's popularity spread across Germany and eventually the world.

When Germans started to immigrate en masse to the United States in the 1840s, Philadelphia was the main point of entry. While many of these Pennsylvania Dutch migrants settled around that area, many others ventured westward to Ohio, Illinois, Missouri, and Texas.[6] Regardless of where these migrants settled, they faced something in common—a lack of lager. Several industrious Germans took it upon themselves to start breweries, whose production was primarily lagers. Such breweries included Pabst (1844, Milwaukee), Anheuser-Busch (1852, St. Louis), Schlitz (1849, Milwaukee), Blatz (1891, Milwaukee), Miller (1855, Milwaukee), Yuengling (1829, Pottsville PA), and Lone Star (1884, San Antonio), among others. While these breweries started small, many of them grew considerably as their market expanded beyond German immigrants to people from many walks of life. The expansion of these German breweries quickly crowded out many smaller breweries, either through acquisition or bankruptcy. The light-colored, low-alcohol, highly carbonated German pilsner offered a different alternative to darker, flatter ales with their higher alcohol content, and non-German beer drinkers by and large quickly switched their preference to lagers. By 1857 lagers were outselling ales, aided no doubt by these German pilsners (Grimm 2018).

Most breweries however, were relativity small, catering largely to neighborhood beer drinkers. This local focus was reinforced due to the lack of refrigeration, which limited brewing and lagering to the cooler months, unless ice blocks were readily available. As a result, breweries located in more northern climes (e.g., Wisconsin, Illinois, etc.) found it easier to brew and preserve lager beer.

Things were about to change for brewers in the United States in the 1870s. Carl von Linde, a German engineer and business administrator who worked for the Spaten brewery in Munich, invented a machine capable of producing ice that allowed for beer to be refrigerated (Wejwar 2011). The adoption of this new technology, coupled with an ever-expanding railway system, allowed lagers to be brewed year round and to be shipped in refrigerated railcars to markets across the United States. The result was explosive growth for many of these German breweries. Today lager is still the most popular beer type in the United States, comprising nearly 80 percent of all beer sold (Notte 2015).

A map of breweries that opened between 1840 and 1910 (fig. 18) shows

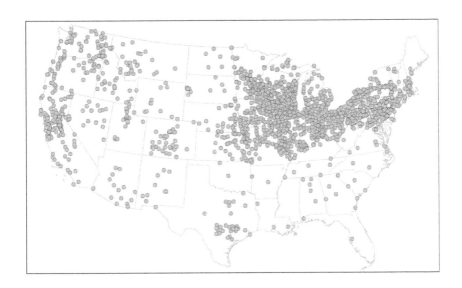

FIG. 18. U.S. brewery openings, 1840–1910. Created by Sam Batzli, 2014.

the peak decades of German migration to the United States (Batzli 2014). During this period more than five million Germans immigrated, particularly to the midwestern cities of Milwaukee, Chicago, Cincinnati, and St. Louis. As a result, distinct geographic patterns emerged, with the vast majority of breweries favoring these cities heavily populated by German immigrants. Another cluster of breweries can also be seen near the San Antonio area in Texas, where those of German ancestry dominated most of the state's urban areas during this period (Jordan 1969). Farther west, brewery clusters appeared mostly where mining proliferated, which became evident during California's gold rush years (1848–55). As beer from the eastern United States regularly spoiled during transit to western parts of the country, it made economic sense to brew beer locally and efficiently. One such startup was Steam beer (now called California Common), which German brewers originated around 1851 as warm-fermented ales using bottom-fermenting (lager) yeast. Quick-brewed beers that could be consumed within three days became a specialty of many California brewers (St. Clair 1998).

Despite the reappearance or creation of ale styles by thousands of craft brewers and breweries, lagers still dominate the market and are an enduring legacy of the ingenuity, pervasiveness, and business savvy of German immigrants and their breweries. In fact beer had surpassed distilled spirits as the most popular alcoholic beverage in the United States by 1890 (Jordan 1969). Beer drinking rose from 3.4 gallons per capita in 1865 to 20 gallons in the early 1910s (United States Brewers Association 1979). German brewers also founded regional and national "shipping breweries," which depended heavily on railroads to transport product. German breweries also incorporated innovations in pasteurizing, bottling, and transportation before Prohibition, as well as acquiring licenses to make "medicinal" beer and developing canning, packaging, and marketing during Prohibition. After the repeal of the Eighteenth Amendment and the lifting of national Prohibition in 1933, the regional German breweries were poised to dominate the industry far into the twentieth century and beyond.

Irish Migration: Proliferation of the Irish Pub (1840–1900)

Like most European immigrants, the Irish immigrated to America to seek a better life. The early waves of Irish immigration coincided with economic hardships in their homeland (e.g., drought, high taxes, rent increases), although religious persecution and political discontent also spurred emigration from Ireland (Sheridan 1979). Most of the early immigrants were Irish Protestants (predominantly Presbyterians) from Northern Ireland and helped establish education, government, and the foundations of society in early colonial America.

Then came the potato famine, a five-year period (1845–50) in which a fungal blight left many in Ireland without crops to pay rent to the landowners or feed themselves. The hardest hit were Catholics from the southern regions of Ireland, and many of them were unskilled, uneducated, and impoverished. Homelessness, disease, and starvation ravaged Ireland (O'Hara 2001). By the time the famine ended, a million people had died. It is no surprise then that between 1841 and 1850 approximately 46 percent of immigrants to the United States arrived from Ireland (Willcox 1931). Those who survived the Atlantic voyages on the overcrowded and unsanitary "coffin ships" (Laxton 1998) made it with very few possessions in tow. In all, more than four million Irish immigrated between 1840 and 1900.

The Irish public house, now pervasive in most cities in the world, was once a local artifact of the Irish community. The earliest taverns in Ireland date back to the seventh century, with origins in ancient Rome; Romans had set up roadside taverns throughout the empire. In the burgeoning new cities of nineteenth-century America, pubs, inns, and taverns became starting points for new immigrants. Public houses were places not only to drink and commune but to bank, shop, secure temporary lodging, and more. For many Irish immigrants the Irish-owned pubs in Boston and other port-of-entry cities re-created the familiar cultural and social environments they had left behind.

By the end nineteenth century the Irish pub had gained prominence in many urban centers of the United States.[7] The Irish pub found its own niche among the working class; it was a niche that complemented rather than competed with the expansive German-style beer gardens (or beer halls) that were quiet, orderly, and even family friendly, as well as the elite dance halls of the day (Irish Central 2016). The saloons, unlike today's Irish-themed bars and commodified *craic* (Grantham 2009; McGovern 2002), likely consisted of a bar or rail at which men would drink beer or serve themselves shots of whiskey (Sismondo 2014).[8]

But the pubs, saloons, and grog shops weren't without prejudice. Particularly for Irish Catholics, the pub became a means to enter the commercial sector, as owning such an establishment provided employment for members of immediate families and other relatives (Fahey 1990), and for some they were the origins of careers in public service (Reilly 1976). As the number of Irish immigrants rose, so too did the number of applicants for liquor licenses. By 1850 the majority of the twelve hundred or so liquor license holders, mostly in New England, were Irish. By 1880 two-thirds of applicants for licenses in Worcester, Massachusetts, were Irish, with more than half of saloon owners being Irish or of Irish descent (Cohen 2004). As for breweries, however, the majority of those that opened during the Irish potato famine years were in midwestern cities like Chicago that were growing rapidly due to the large numbers of German and Irish Catholic immigrants. This geographic pattern is quite evident in figure 19.

Dissenting voices grew louder in response to the large numbers of immigrants, and Irish Catholics and Germans were favorite targets. Temperance supporters and nativists lamented the high number of Irish-owned bars, taverns, and grog shops, as well as the numerous unlicensed shebeens, which were illegal bars run by Irish women within the confines of their homes (Sismondo 2014). The hostility toward immigrants, particularly Irish Catholics and their penchant for owning, distributing, and consuming alcohol, came to a head in Maine when the state began prohibiting alcohol sales.[9] Blue laws, which forbade the sale of alcohol on Sundays, were implemented in some places, sometimes statewide. For many immigrants Sunday was the only day of the week they were not working and thus the only day they could drink alcohol (Sismondo 2014).

Guinness beer, one of the best-known beer brands from Ireland, was first documented in the United States in 1817, thanks to an order placed by a man named John Heavy in South Carolina. And while many love the Irish dry stout, the overall popularity of Irish-style beers has paled in comparison to that of American and German lager styles, including the widely popular post-Prohibition American adjunct lager (Dighe 2016).[10] Nevertheless, the brand grew in popularity, based on a series of marketing campaigns starting in the late 1920s, the creation of the Guinness Book of World Records in the 1950s, and the pairing of Guinness beer with the Irish Pub Company to generate a global Irish pub diaspora beginning in the 1990s.

Saint Patrick's Day, known in the United States for its parades, revelry, and beer drinking, originated in Ireland as a Roman Catholic feast day to honor the patron saint of Ireland, who is credited with bringing Christianity to the island. How it came to the United States was by way of the Irish soldiers in the British army of the eighteenth century and

FIG. 19. U.S. brewery openings, 1845–50. Created by Sam Batzli, 2014.

later the mass immigration of Irish in the nineteenth century (Little 2018). The celebration of Saint Patrick's Day, complete with festivities, speeches, and sermons, served as "reflections of Irish memory and identity" (Moss 1995, 130). Municipalities where large populations of Irish immigrants lived began backing celebrations, and by the late twentieth century the holiday had become so secularized that it was easily adopted into mainstream American culture. Guinness has taken center stage in this celebration, becoming the emblematic beer of Saint Patrick's Day for most Americans. This is a bit ironic, historically speaking, as Guinness was once a company so aligned with the Protestants that any employee who married a Catholic would be asked to resign (Hershberger 2018).

Mexican Migration: The Return of European Preferences (1986–2017)

Beer in Mexico was once scarce, with most Mexicans drinking traditional fermented beverages, such as *pulque*, the milk-colored, sour-tasting drink made from the fermented sap of the agave plant. But beer from Europe became central to Mexico's people, in terms of industry and as the preferred beverage. The rise of the European-led beer culture in Mexico is a story of wars, empires, political boundaries, beer campaigns, Prohibition, and mass migration.

The introduction of European beer and brewing to Mexico is one that begins in Europe and in particular Austria. In 1841, a year prior to the invention of pilsner, Anton Dreher developed a new style of beer in Vienna, a result of his knowledge of Germany's lagering techniques and England's new kilning technologies. While Austrian brewers were not constrained by the German beer purity law (Reinheitsgebot) and its three-ingredient restriction (water, barley, and hops), the styles were nonetheless similar to those of nearby Bavaria.[11] With time the Vienna lager style waned in popularity, due in part to the rise of the well-branded, malty sweet Austrian märzen.[12]

Vienna lager, however, would find a new home in Mexico. With the support of the French army, Maximilian, the younger brother of the Austrian emperor Francis Joseph, ousted the new Mexican president Benito Juárez in 1864 and declared himself emperor of Mexico. While the empire lasted only three years, it prompted Germans and Austrians to establish themselves in various settlements.[13] The emperor's predilection for Vienna lager brought the (likely) golden yellow, full-bodied, low-bitterness style to Mexico (for more on the historic style characteristics, see Leyser and Heiss 1900), and by the 1920s the style had become synonymous with beer brewed in Mexico.

In addition, the diaspora of Germans and Austrians was not limited to the United States; migration from Germany was evident in settlements in Texas, first under Spanish rule and then Mexican. Brewers from Germany, Switzerland, and Alsace (the region of France bordering Germany) established breweries in newly formed settlements of immigrants. By way of war, land cession, and various purchases from foreign powers, the American West quickly became a series of territories within the United States. By this manner many of these Mexican breweries soon

Texas 1838-1848
Texas territory in dispute
Territory lost by Mexico in 1846
Mexico after 1848

FIG. 20. U.S.-Mexico boundary changes, 1840s. Map produced by authors with data from J. Rickard, "Mexican War (May 1846–February 1848)," February 1, 2001, http://www.historyofwar.org/articles/wars_mexican.html.

became U.S. breweries, particularly as the result of the annexation of Texas (1845), the U.S.-Mexico War (1846–48), and the Gadsden Purchase in 1853 (fig. 20).

While Mexico claims the first European-style brewery in North America, opened by royal decree and Don Alonso de Herrera during the Spanish conquest in the 1540s (Gauss and Beatty 2014), it wasn't until the 1890s that many of today's leading breweries emerged. The establishment and growth of these breweries coincided, perhaps unsurprisingly, with the growth of cities and urban areas in Mexico (Gauss and Beatty 2014). When passage of the Eighteenth Amendment instituted Prohibition across the whole of the United States in 1920 (until its repeal in 1933), Americans flocked to Mexico, increasing demand and solidifying European lagers as favorite Mexican beer styles. The once popular pulque was defamed by beer campaigns castigating the "unsanitary conditions" that produced the indigenous drink, further promoting the consumption (and thereby production) of European-style beer in Mexico. Today Mexico ranks as the top country for beer exports, with Mexican lagers accounting for the vast majority of beer produced and consumed (Euromonitor 2017).

While migrations from Mexico to the U.S. states on the border have been ongoing since the establishment of the boundary in the 1850s, a surge of migration occurred first between 1900 and 1930, as political exiles of the Mexican Revolution and individuals from rural communities sought refuge and stayed to seek new economic opportunities. A more recent surge of Mexican immigration occurred after 1965, when U.S. immigration laws emphasizing family reunification replaced the previous national origins quota system. The Immigration Reform and Control Act (1986) legalized 2.3 million Mexican immigrants in exchange for mandated employer sanctions for anyone who hired illegal aliens.[14] By 2015 the foreign-born U.S. population included 11.6 million (26 percent) immigrants from Mexico (López and Bialik 2017). Mexican immigrants account for more than half of the foreign-born population in the states of New Mexico (69 percent), Texas (55 percent), Arizona (55 percent), and Idaho (51 percent).[15]

In the year that the term "craft brewery" was likely coined (Cottone 1986), only 124 breweries were operating within the United States (Brewers Association 2017); just after the peak of Mexican immigration in 2011, that figure had grown to 2,047 (see fig. 21 for locations). Breweries established during this time were by and large "craft breweries," as defined by the Brewers Association (2017); these are located predominantly along the coasts, in Colorado, and in the historically German-settled cities of the Midwest and Texas. As the number of Mexican migrants increased, so

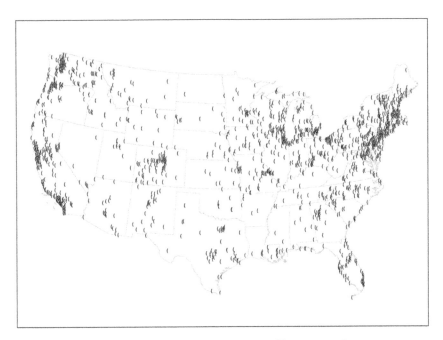

FIG. 21. U.S. brewery openings, 1986–2011. Created by Sam Batzli, 2014.

too has their role in the beer industry. This is evident with the increased number of Mexican Americans founding breweries, such as Keith Villa for Blue Moon in Denver, Diego Benitez for Progress Brewery in Los Angeles, and cofounder Isaac Showaki for Rabbit Cervecería in Illinois, among others. While most breweries owned or inspired by Mexicans or Mexican Americans were established after 2011, their locations are predominantly in cities with higher Mexican immigrant populations.

Although Mexican immigrants by no means constitute the majority of craft brewery owners, it is likely that the influx of Mexicans to the United States has promoted the popularity of Mexican beers, especially as global beer conglomerates market to Mexican-born immigrants, their descendants, and multicultural millennials (Kell 2015). Indeed two-thirds of America's imported beer is Mexican, which means that one of every nine bottles of beer in the United States comes from Mexico (Pomranz 2018). As a result, many craft brewers in the United States have taken notice of this trend and responded by producing American-made Mexican lagers. Such versions are commonly characterized by toasty or bready maltiness, light, crisp mouthfeel, and low bitterness (see, e.g., Weaver 2017).

One of the best examples of how Mexican culture has shaped the American beerscape is the celebration of Cinco de Mayo, an annual holiday that commemorates the Mexican victory over the French in the Battle of Puebla in 1862. While the day is observed and rightfully celebrated in the Mexican state of Puebla, the U.S. version has largely been co-opted by beer companies, becoming a day of carousing. According to José Alamillo (2009), the advent of the modern-day Cinco de Mayo celebration in the United States can be traced to the early 1980s, when Anheuser-Busch and Miller looked for ways to reach the U.S. Hispanic market. The brewing conglomerates hit upon sponsoring and actively marketing Cinco de Mayo festivities in Hispanic communities, and soon other beer importers, such as Heineken (Dos Equis and Tecate) and Gambrinus (Corona and Modelo), followed suit. So successful were their campaigns that Americans now spend more on beer for Cinco de Mayo celebrations than they do for Saint Patrick's Day or the Super Bowl (Tyler and Huffman 2014). Moreover, Cinco de Mayo festivities have expanded from Hispanic communities; many Americans embrace Cinco de Mayo as a national day of drinking.

While Mexican immigrants are the most recent of the migrant groups addressed here, they have had a significant impact on the American beerscape. Their penchant for Vienna lagers has led to a significant growth in the number of breweries producing the style. Moreover, their celebration of Cinco de Mayo was seen as an opportunity for the beer industry to market beer to Mexican immigrants. This led to the eventual co-optation of the celebration, which was so successful that the only holidays for which more beer is sold are July 4, Memorial Day, Labor Day, and the Christmas holidays (National Beer Wholesalers Association 2016). Like the other three migrant groups, Mexican immigrants have left a big impact on America's beer landscape.

Conclusion

Since the early 1600s immigration has been one of the major drivers influencing the beer landscape of the United States. In this chapter we have addressed the impacts from four significant migration periods and shown how immigrants have shaped the U.S. beerscape in terms of brewery locations, beer styles, and the transculturation of beer. The English brought forth the tradition of brewing beer in America, particularly in an industrial sense, and incorporated beer production and consumption into daily life. The taverns established by the British in the colonies would spread across the landscape, ultimately providing meeting places for news to be exchanged, gossip to be spread, legal rulings to be made, and rebellions to be planned. Agrarian developments arising from the need to grow barley and hops for the production of beer were largely spearheaded by British settlers.

The Germans continued the English tradition of brewing, but on a much larger and broader scale. German-style lagers, arising in the Midwest, would eventually spread across the land and overtake British ales. The size of German breweries helped them to shape the fermented landscape through acquisitions and allowed many of these breweries to survive the "grand experiment" known as Prohibition. Breweries were no longer a local phenomenon, but through the implementation of refrigerated railcars—a German invention—many German breweries became regional ones and in some cases national in scope. Innovations such as pasteurization and bottling were readily incorporated into the production process by German-founded breweries and were gradually adopted by other breweries.

The Irish continued the establishment of public houses, which functioned as community centers for the exchange of news and gossip and as sites for an assortment of other activities, such as judicial hearings, banking, and rest. By the 1900s there wasn't an urban center in the United States without an Irish pub. Today Irish pubs continue to be an important fixture in U.S. society, especially as a locale for social discourse and exchange. With the commodification of Irish craic and the nationalization of Saint Patrick's Day, the Irish drinking experience has been further embedded into America's fermented landscapes. The popularity of European-style lagers, such as the Vienna lager from Mexico, was initially introduced to the United States as a result of Prohibition; Mexican breweries provided Vienna lager (and other beer styles) to Americans willing to cross the border in search of beer during Prohibition. In addition, Mexican migrants to the United States brought with them their penchant for Vienna lagers (e.g., Negra Modelo, Dos Equis Amber), which quickly became popular in the U.S. Southwest. Today many American breweries are capitalizing on the popularity of these lagers by creating their own interpretation of the style, ensuring that Vienna lager will live on.

America's beer landscapes are a product of migrants bringing their know-how, preferences, and cultures from their country of origin to their new homeland. Each group faced different physical environments, cultural norms, and social adversities upon their arrival, yet in spite of (or perhaps because of) these differing conditions, all four groups of immigrants were able to make a lasting impact on the U.S. beerscape and contribute to the ever-evolving fermented landscapes of the United States.

Notes

1. Bruman (2000) argues that no alcoholic beverages were brewed north of what is now Arizona and New Mexico.
2. See chapter 1 of this volume for a conceptualization of fermented landscapes.
3. The Dutch reclaimed New York briefly in 1673.
4. See chapter 2 for a similar discussion of the layered values of spaces like taverns and pubs.
5. The verb "lager" comes from the German word *lagern*, which means to store.
6. The term "Pennsylvania Dutch" refers to German settlers in the state, as "Dutch" is the anglicized word for *deutsch*, or German.
7. Established in 1854, McSorley's Old Ale House, the oldest Irish bar in New York City, went one hundred years before women were allowed to enter, which happened as the result of a lawsuit.

8. Most Irish pubs today are not authentic Irish pubs but rather a mass-produced commodification of craic, which loosely translates to having a good time (e.g., drinking, banter, gossip, debauchery) in the company of others.

9. It should be noted that alcohol consumption was never prohibited; rather, it was the production and distribution of alcohol, which specifically discriminated against those who made beer (Germans) and those who sold it (Irish).

10. An adjunct lager is a lager made with ingredients beyond the basic barley, hops, water, and yeast. Common adjuncts include rice and corn as part of the grain bill.

11. The Reinheitsgebot did not mention yeast, but brewers well understood its role. Duke Wilhelm of Bavaria likely omitted it from that famous beer decree he issued, as he mentions only the ingredients that stayed in the beer, unlike yeast, which was always removed and added to the next batch of beer (Alworth 2016).

12. Austrian märzen came with a catchy slogan: "Brewed the Vienna way" (Dornbusch 2011).

13. Mexican liberals, with the support of the post—Civil War U.S. government, eventually ousted and executed Maximilian.

14. Ironically, these sanctions were rarely enforced and helped to increase, rather than decrease, the number of unauthorized workers in the United States.

15. Apart from Idaho, the remaining states were part of the original Mexican territory ceded to the United States after gaining statehood (Texas) or as a result of the U.S.-Mexico War.

References

Alamillo, José M. 2009. "Cinco de Mayo Inc.: Reinterpreting Latino Culture into a Commercial Holiday." *Studies in Symbolic Interaction* 33:217–38.

Alworth, Jeff. 2016. "Attempting to Understand the Reinheitsgebot: 500 Years of Brewing Law." *All about Beer* 37 (1). http://allaboutbeer.com/article/happy-birthday-reinheitsgebot/.

Batzli, Sam. 2014. "Mapping United States Breweries 1612 to 2011." In *The Geography of Beer: Regions, Environment, and Societies*, edited by Mark W. Patterson and Nancy Hoalst-Pullen, 31–43. Dordrecht: Springer. https://doi.org/10.1007/978-94-007-7787-3.

Brewers Association. 2017. "Numbers of Breweries: Historical U.S. Brewery Count." Accessed February 28, 2018. https://www.brewersassociation.org/statistics/number-of-breweries.

Brooks, Rebecca Beatrice. 2017. "The Great Puritan Migration." *History of Massachusetts Blog*, May 24, 2017. http://historyofmassachusetts.org/the-great-puritan-migration/.

Bruman, Henry J. 2000. *Alcohol in Ancient Mexico*. Salt Lake City: University of Utah Press.

Chapelle, Francis H. 2005. *Wellsprings: A Natural History of Bottled Spring Waters*. New Brunswick NJ: Rutgers University Press.

Cohen, Bruce. 2004. "Ethnic Catholicism and Craft Unionism in Worcester, Massachusetts, 1887–1920: A Mixed Story." *Historical Journal of Massachusetts* 32 (2). http://www.westfield.ma.edu/mhj/pdfs/Cohen%20summer%202004%20combined.pdf.

Cottone, Vince. 1986. *Good Beer Guide Breweries and Pubs of the Pacific Northwest*. Seattle: Homestead Book Company.

DHS (Department of Homeland Security). 2008. *2008 Yearbook of Immigration Statistics*. Washington DC: Office of Immigration Statistics. https://www.dhs.gov/xlibrary/assets/statistics/yearbook/2008/ois_yb_2008.pdf.

Devitt, Terry. 2016. "Lessons of Lager: Yeast Origin Becomes a Complex Tale." University of Wisconsin—Madison News, July 6, 2016. https://news.wisc.edu/lessons-of-lager-yeast-origin-becomes-a-complex-tale/.

Dighe, Ranjit S. 2016. "A Taste for Temperance: How American Beer Got to Be So Bland." *Business History* 58 (5): 752–84.

Dornbusch, Horst. 2011. "Vienna Lager." In *The Oxford Companion to Beer*, edited by Garrett Oliver, 816–17. Oxford: Oxford University Press.

Euromonitor International. 2017. "Beer in Mexico." Accessed May 20, 2018. http://www.euromonitor.com/beer-in-mexico/report.

Fahey, Tony. 1990. "Catholicism and Industrial Society in Ireland." *Proceedings of the British Academy* 79:241–63.

Gauss, Susan M., and Edward Beatty. 2014. "The World's Beer: The Historical Geography of Brewing in Mexico." In *The Geography of Beer: Regions, Environment, and Societies*, edited by Mark W. Patterson and Nancy Hoalst-Pullen, 57–65. Dordrecht: Springer. https://doi.org/10.1007/978-94-007-7787-3.

Grantham, Bill. 2009. "*Craic* in a Box: Commodifying and Exporting the Irish Pub." *Continuum: Journal of Media & Cultural Studies* 23 (2): 257–67. https://doi.org/10.1080/10304310802710553.

Grimm, Lisa. 2018. "Beer History: German-American Brewers before Prohibition." *Serious Eats*. Accessed May 20, 2018. https://drinks.seriouseats.com/2011/10/beer-history-german-american-brewers-before-prohibition.html.

Heisel, S. 2011. "American Hops, History." In *The Oxford Companion to Beer*, edited by Garrett Oliver, 43. Oxford: Oxford University Press.

Hershberger, Matt. 2018. "How Guinness Went from Anti-Catholic to King of St. Patrick's." *Eat Sip Trip*. Accessed March 17, 2018. https://mix.com/!fLQdfMwt:how-guinness-went-from-anti-catholic-to-king-of-st.-patrick's.

Irish Central. 2016. "From the Famine to Five Points—History of Irish Pubs in the USA." September 12, 2016. https://www.irishcentral.com/news/community/from-the-famine-to-five-points-history-of-irish-pubs-in-the-usa.

Jordan, Terry G. 1969. "Population Origins in Texas, 1850." *Geographical Review* 59 (1): 83–103.

Kell, John. 2015. "Why Mexican Beers Continue to Outperform the Industry." *Fortune*, December 30, 2015. http://fortune.com/2015/12/30/mexican-beers-strong-sellers/.

Keppel, Patricia. 2018. "The Story behind the Craft: Discovering Virginia's Beer History." *Travel Ideas and Stories: Virginia's Travel Blog.* Last updated March 26, 2018. https://blog.virginia.org/2017/07/virginias-beer-history/.

Kobler, John. 1973. *Ardent Spirits: The Rise and Fall of Prohibition.* New York: G. P. Putnam's Sons.

Krakowski, Adam. 2014. "A Bitter Past: Hop Farming in Nineteenth-Century Vermont." *Vermont History* 82 (2): 91–105.

La Barre, W. 1938. "Native American Beers." *American Anthropologist* 40 (2): 224–34. https://doi.org/10.1525/aa.1938.40.2.02a00040.

Laxton, Edward. 1998. *The Famine Ships: The Irish Exodus to America.* New York: Henry Holt.

Leyser, Emil, and Philipp Heiss. 1900. *Die Malz-und Bierbereitung: Ein Handbuch zum Selbstunterricht für Praktiker sowie zum Gebrauche an Brauerschulen.* Stuttgart: M. Waag.

Libkind, Diego, Chris Todd Hittinger, Elisabete Valério, Carla Gonçalves, Jim Dover, Mark Johnston, Paula Gonçalves, and José Paulo Sampaio. 2011. "Microbe Domestication and the Identification of the Wild Genetic Stock of Lager-Brewing Yeast." *Proceedings of the National Academy of Sciences* 108 (35): 14539–44. https://doi.org/10.1073/pnas.1105430108.

Little, Becky. 2018. "How America, Not Ireland, Made St. Patrick's Day as We Know It." *National Geographic*, March 14, 2018. http://www.nationalgeographic.com.au/people/how-america-not-ireland-made-st-patricks-day-as-we-know-it.aspx.

López, Gustavo, and Kristen Bialik. 2017. "Key Finding about U.S. Immigrants: Fact Tank—Our Lives in Numbers." Pew Research Center. Accessed May 20, 2018. http://www.pewresearch.org/fact-tank/2017/05/03/key-findings-about-u-s-immigrants/#.

Lundmark, Michael. 2015. "How Beer Single-Handedly Saved the State of Georgia." *Suwanee Magazine*, May 2015. https://suwaneemagazine.com/whats-brewing-3/.

McGovern, Mark. 2002. "'The "Craic" Market': Irish Theme Bars and the Commodification of Irishness in Contemporary Britain." *Irish Journal of Sociology* 11 (2): 77–98. https://doi.org/10.1177/079160350201100205.

Moss, Kenneth. 1995. "St. Patrick's Day Celebrations and the Formation of Irish-American Identity, 1845–1875." *Journal of Social History* 29 (1): 125–48. http://www.jstor.org/stable/3788712.

National Beer Wholesalers Association. 2016. "Fourth of July: Where It Ranks among Beer Holidays." NBWA. Accessed May 24, 2018. https://www.nbwa.org/resources/fourth-july-where-it-ranks-among-beer-holidays.

Notte, Jason. 2015. "These 11 Brewers Make Over 90% of All U.S. Beer." Market Watch, July 28, 2015. https://www.marketwatch.com/story/these-11-brewers-make-over-90-of-all-us-beer-2015-07-27.

O'Brien, J. 2015. "The Time When Americans Drank All Day Long." *BBC News*, March 9, 2015. http://www.bbc.com/news/magazine-31741615.

O'Hara, Megan. 2001. *Irish Immigrants: 1840–1920 (Coming to America)*. New York: Capstone Press.

Patterson, Mark W., and Nancy Hoalst-Pullen, eds. 2014. *The Geography of Beer: Regions, Environment, and Societies*. Dordrecht: Springer. https://doi.org/10.1007/978-94-007-7787-3

Phillips, Rod. 2014. *Alcohol: A History*. Chapel Hill: University of North Carolina Press.

Pomranz, Mike. 2018. "Two-Thirds of America's Imported Beer Is Mexican." *Food and Wine*. Updated February 16, 2018. https://www.foodandwine.com/news/mexican-beer-usa.

Reilly, Joseph. 1976. "The American Bar and the Irish Pub: A Study in Comparisons and Contrasts." *Journal of Popular Culture* 10 (3): 571–78. https://doi.org/10.1111/j.0022-3840.1976.1003_571.x.

Rickard, J. 2001. "Mexican War (May 1846–February 1848)." Accessed April 25, 2018. http://www.historyofwar.org/articles/wars_mexican.html.

Sheridan, Peter B. 1979. "The Protestant Irish Heritage in America." *Études Irlandaises* (4):167–76. https://doi.org/10.3406/irlan.1979.2597.

Sismondo, Christine. 2014. *America Walks into a Bar: A Spirited History of Taverns and Saloons, Speakeasies and Grog Shops*. Oxford: Oxford University Press.

Smith, Daniel S. 1972. "The Demographic History of Colonial New England." *Journal of Economic History* 32 (1): 165–83.

Smith, Gregg. 1998. *Beer in America: The Early Years, 1587–1840; Beer's Role in the Settling of America and the Birth of a Nation*. Denver: Brewers Publications.

St. Clair, David J. 1998. "The Gold Rush and the Beginnings of California Industry." *California History* 77 (4): 185–208. https://doi.org/10.2307/25462514.

Stinehour, Zach. 2016. *An American Beer Trail*. Parker CO: Outskirts Press.

Tyler, Jeff, and Shea Huffman. 2014. "The History of the Marketing of Cinco de Mayo." Marketplace, May 2, 2014. https://www.marketplace.org/2014/05/02/business/history-marketing-cinco-de-mayo.

United States Brewers Association. 1979. *1979 Brewers Almanac*. Washington DC: United States Brewers Association.

Vriesekoop, Frank, Moritz Krahl, Barry Hucker, and Garry Menz. 2012. "125th Anniversary Review: Bacteria in Brewing; The Good, the Bad and the Ugly." *Journal of the Institute of Brewing* 118 (4): 335–45. https://onlinelibrary.wiley.com/doi/epdf/10.1002/jib.49.

Weaver, Ken. 2017. "Mexican Style Lagers: Mas por Menos." *All about Beer*, September 17, 2017. http://allaboutbeer.com/article/mexican-style-lagers/.

Wejwar, Sepp. 2011. "Austria." In *The Oxford Companion to Beer*, edited by Garrett Oliver, 73–75. Oxford: Oxford University Press.

Willcox, Walter F. 1931. "Immigration into the United States." In *International Migrations, Volume II: Interpretations*, edited by Walter F. Willcox, 83–122. Washington DC: National Bureau of Economic Research.

The *Goût du Terroir* and Culinary Culture of Bloody Mary Cocktails in the United States

7

Paul Zunkel

Food is an integral component in the overall experience of a place (Birch 2017). Food transmits a plethora of cultural information; it constitutes the basis for a culinary culture specific to a place or region. This food-based cultural information is absorbed, processed, and conveyed by locals and tourists seeking authentic regional or local experiences (Metro-Roland 2013). Investigating food and the place where it was created provides opportunities to explore both the production and consumption of food without privileging one over the other (Goodman 2016).

The terms *terroir*, derived from the Latin word for "earth," and *goût du terroir*, which is French for "taste of the earth," were originally used to describe the relationship between wine and the geographic location where that wine was produced, as well as local environmental factors such as soil, topography, and climate (Trubek 2008; Jacobsen 2010). The association of taste, place, and quality with the concept of terroir is a more recent development (Trubek 2008). While credit is frequently given to France for the development of terroir, examples can be found throughout the world and even in the United States. Rowan Jacobsen (2010) describes several examples of items in North America that have a particular terroir, including salmon in Alaska, chocolate in Mexico, apples in Washington's Yakima Valley, wild mushrooms in Quebec, and cheese in Vermont. According to Goodman (2016, 258), when studying food, it is impossible to disregard the notions of culture, space, economy, and politics.

Jacobsen (2010, 5) explains how Cajun gumbo is an example of a dish having a particular terroir, since gumbo evolved to incorporate the best of what the land and bayous had to offer, including regional or local ingredients such as crayfish and sassafras leaves. The Bloody Mary cocktail drink is similar in this regard. Bloody Marys are a unique sociocultural example of both global and local symbolic fermentation because their ingredients are found and produced both globally and locally. Certain ingredients, such as tomato juice, lemons or lemon juice, vodka and other spirits, salt, and pepper, may be produced at locations across the United States or the world. However, garnishes and toppings such as venison sticks, sugarcane, hot sauce, jalapeño peppers, and Old Bay seasoning are typically produced regionally or locally and offer a specific lens with which to examine the production and consumption of a product that incorporates a fermented material.

The purpose of this study is to further elucidate the relationship between identity and place, as well as production and consumption, by ethnographically examining the Bloody Mary cocktail drink. The objectives guiding this study are threefold: (1) to document the history of the Bloody Mary, (2) to examine how Bloody Mary garnishes and condiments have become icons of regional and local culture, and (3) to discuss regional and local Bloody Mary recipes while focusing on the culinary culture and goût du terroir of condiments and garnishes. The author's previous experiences in travel and with food and drink history provide examples that show how Bloody Marys, along with garnishes and condiments, offer a message regarding the identity, ceremony, habits, and customs of a regional or local society. This study places examples of terroir, goût du terroir, and the culinary culture of alcohol products within the larger framework of symbolic fermentation, which scholars have yet to address at any length in the literature.

FIG. 22. The "Bloody Beast" Bloody Mary at Sobelman's Pub & Grill, Milwaukee, Wisconsin. Image courtesy Sobelman's Pub & Grill. Used with permission.

The Bloody Mary

Examples of eccentric Bloody Mary concoctions, including an elaborate example more than twelve inches (0.3 m) tall (fig. 22), can be found throughout the United States and the world (Birch 2017). However, the basic Bloody Mary recipe is quite simple. For example, the International Bartenders Association (IBA) issues standards for classic and contemporary cocktails, including the ingredients needed, directions on how to build the drink, the appropriate glassware, and garnishes.

The basic recipe is as follows:

1½ ounces vodka

3 ounces tomato juice

½ ounce lemon juice

2 to 3 dashes of Worcestershire sauce

Tabasco

Celery salt

Pepper

Stir gently, pour all ingredients into highball glass. Garnish with celery and lemon wedge (optional). (IBA 2017)

The recipe, while uninteresting to many consumers, is important because bartenders, that is, the producers of the drink, take this recipe, deviate from it, and incorporate regional or local liquid ingredients, condiments, or garnishes to create a product that is appealing to consumers. The incorporation of regional and local ingredients gives a sense of place to the cocktail and makes the Bloody Mary an example of culinary culture and of a drink with its own goût du terroir. This study will discuss and document local and regional variations of the Bloody Mary cocktail based on the IBA's basic Bloody Mary recipe.

History of the Bloody Mary

Beginning in 1920 and throughout much of that decade, elite bartenders from the United States relocated to Europe after the United States passed the Eighteenth Amendment, which outlawed the production, transport, and sale of alcohol. American bartenders escaping Prohibition and a newly "dry" America brought with them all the comforts of their former establishments, including canned tomato juice (Petro 2014). At the same time, Russia was becoming a laboratory for the ideas of Karl Marx and Vladimir Lenin, and some Russians were escaping their homeland and relocating to western Europe. These Russians brought with them a bold, determined spirit and another spirit called vodka. Vodka was not a popular spirit outside of Russia, Poland, and Scandinavian countries at that time, and its arrival in Europe, and in places such as Paris, gave bartenders, both European and American, a fresh ingredient for experimentation.

While the exact origin of the Bloody Mary is uncertain, the Parisian bartender Fernand Petiot is frequently credited with inventing the earliest version, at Harry's New York Bar in Paris in the early 1920s (Bartels 2017; Birch 2017; Sutcliffe 2017). The original cocktail is said to have consisted of only vodka and canned tomato juice from the United States (Park 2008; Chazan 2011). Debate remains regarding the origin of the drink's name. According to Harry's New York Bar manager Alain Da Silva, Petiot took the name from a customer named Mary who worked at a cabaret club called the Bucket of Blood (Chazan 2011; Sutcliffe 2017). Others believe the Bloody Mary refers to Queen Mary I of England, with the drink's red color being a direct reference to her brutal persecution of Protestants during the English Reformation (Haigh 1987).

The 21 Club in New York City previously claimed that the Bloody Mary was invented in the 1920s by a bartender named Henry Zbikiewicz (Smith 2007, 55). George Jessel, a comedian, actor, and frequenter of the 21 Club, claimed that he invented the Bloody Mary in Palm Beach, Florida, after going on a wild drinking spree and needing some "hair of the dog" (Smith 2007; Birch 2017). The phrase "hair of the dog" is an expression for an alcoholic drink taken to cure a hangover and is shortened from the phrase "a hair of the dog that bit you" (Paulsen 1961). The phrase originates from an old belief that a person bitten by a rabid dog could be cured of rabies by taking a potion containing some of the dog's hair (Paulsen 1961; Calder 1997; Spence 2003). The correlation suggests that, although alcohol may be to blame for the hangover, as the dog is for the attack, a smaller portion of alcohol will act as a cure (Calder 1997). In the 1950s Jessel partnered with Smirnoff to promote the cocktail as his own creation (Sutcliffe 2017).

One boost to the Bloody Mary's popularity was the rise of brunch, which originated in England in the late nineteenth century and became popular in the United States in the 1930s (Rombauer, Becker, and Becker

2001, 8). The Bloody Mary evolved into the well-known brunch staple after Petiot left Paris in 1925 and came to the United States. In 1933, after several years in Ohio, Petiot became the head bartender for the King Cole Bar at the St. Regis Hotel in New York City (Sutcliffe 2017). In 1934 the Russian prince and businessman Serge Obolensky came to New York and ordered a Bloody Mary at the King Cole but requested one with more spice and flavor. Petiot modified the original mixture by adding salt, black pepper, cayenne pepper, Worcestershire sauce, and lemon juice and then garnished the drink with a celery stick. According to a 1964 *New Yorker* article,

> "I initiated the Bloody Mary of today," he [Petiot] told us. "Jessel said he created it, but it was really nothing but vodka and tomato juice when I took it over. I cover the bottom of the shaker with four large dashes of salt, two dashes of black pepper, two dashes of cayenne pepper, and a layer of Worcestershire sauce; I then add a dash of lemon juice and some cracked ice, put in two ounces of vodka and two ounces of thick tomato juice, shake, strain, and pour. We serve a hundred to a hundred and fifty Bloody Marys a day here in the King Cole Room and in the other restaurants and the banquet rooms." (quoted in Sutcliffe 2017)

Vincent Astor, then owner of the St. Regis Hotel, renamed the drink the Red Snapper because he believed the name "Bloody Mary" was too vulgar and not befitting of such a prestigious hotel (Birch 2017; Sutcliffe 2017; "Bloody Mary" 2017). The cocktail was renamed the Red Hammer in *Life* magazine in 1942 and consisted of tomato juice, vodka, and lemon juice (*Life* 1942). Around the same time, David Dodge (1942) suggested adding salt as a basic ingredient. In the 1960s the first commercial mix, Mr. and Mrs. T Bloody Mary, was offered to the public (Petro 2014).

Local Food Culture

The American philosopher Charles Sanders Peirce used semiotic theory to look beyond mere symbolic relationships between words and concepts to explain the ways in which humans interpret the world of physical objects that surround them (Houser and Kloesel 1992). A simple explanation of semiotic theory is that signs offer information that provides for denotation (i.e., the literal or primary meaning) and connotation (i.e., an idea or feeling in addition to its literal or primary meaning) and that increasing use of these symbols becomes inexorably linked to customs and culture (Houser and Kloesel 1992).

Food, as well as the act of eating, drinking, tasting, and digesting, is intimately linked to bodies, places, subjectivities, and sociospatial relations (Del Casino 2015, 804). What people eat, how they eat it, and what it means to them is wrapped up in broader relations and processes that make the act of engaging with food complex and spatially differentiated (Probyn 2000). For example, consider the dumpling. Jennifer Jordan (2008, 109) notes that dumplings can have multiple meanings to an individual and can be symbols of regions or nations, objects of fading nostalgia, or elements of active entrepreneurial campaigns to boost the economies and external identities of particular regions. In France, terroir is often associated with an individual's history with a certain place. Local taste, or goût du terroir, is often evoked when an individual wants to remember an experience, explain a memory, or express a sense of identity (Trubek 2008, 51). Local foods can be embedded in long-standing cultures of consumption in which the qualities of the product are consistent with the local notions of taste (Flandrin and Montanari 1999; Montanari 2006; Morgan, Marsden, and Murdoch 2009, 12). Local foods help constitute and define the idea of place, through taste, in the ever-changing global flow of ideas, ingredients, and values (Trubek 2008, 53). Furthermore, needs, desires, and tastes shift over time and space and in relation to food itself (Probyn 2000; Del Casino 2015). Independent local restaurants, inspired by various influences and using local ingredients, create food products and a culinary culture that offers insight into the region or local area, its history, and traditions, and those who eat in such restaurants can better appreciate the quality of place, which in turn creates a distinct destination (Nelson 2005).

The Contributions of Bloody Mary
Culinary Culture to Regional Identity

Bloody Mary goût du terroir and culinary culture can be identified by examining the local ingredients used in the cocktail. Recipes incorporate regional or local spirits, beer, and garnishes that help express a cultural theme or idea that incorporates the culture, history, and traditions of that place or region. The following subsections detail regional Bloody Mary variations and their links to culinary culture and goût du terroir in the United States.

The North and Midwest

Bloody Marys in the North and Midwest are frequently served with a beer chaser. The practice of serving a beer chaser, or "snit," as it is commonly referred to by locals, is a regional practice in Minnesota, Wisconsin, Iowa, Michigan, and Illinois (Henzl 2016). A snit typically contains three to five ounces of beer and is served in a highball or juice glass (Cassel 2018) (fig. 23). While the origin of this tradition is unknown, there are several possible explanations. One focuses on an early version of the Bloody Mary, made with tomato juice and beer. A Russian vodka shortage in the 1950s is suspected to be the catalyst for bartenders substituting locally produced beers, such as Hamm's or Grain Belt, for the traditional vodka (Brunchkateers 2012; Cassel 2018). At the end of the vodka shortage, the beer continued to be served on the side—to break down the spiciness and for "good measure" (Olson 2009). A second idea postulates that the tradition of serving a beer chaser is because of the large number of breweries throughout the region. For many of the German, Irish, and Scandinavian migrants who settled in the North and Midwest, the cultural practice of drinking was a common occurrence on the factory line, at lunch, and on into the evening (Olson 2009; Cassel 2018). Bloody Marys from this region often include locally produced garnishes such as venison or jerky sticks, summer sausage, bacon, local mushrooms, and cheese cubes or curds, all of which suggest farming, rural life, and outdoor activities (Olson 2009). Here is a Chicago establishment's recipe:

1½ ounces vodka

4⅔-ounce can of tomato juice

¼ teaspoon celery salt

¼ tablespoon black pepper

¼ tablespoon horseradish

¼ tablespoon garlic powder

⅓ ounce Worcestershire sauce

⅛ ounce lime juice

⅛ ounce lemon juice

¼ ounce simple syrup

Garnish:
Carr Valley Marisa sheep's milk cheese from Wisconsin
Spanish chorizo or meat
Whiskey sour pickle spear

Combine ingredients, stir, and garnish. (LaMorte 2016)

The East and Northeast

Bloody Mary recipes from the East and Northeast also incorporate regional ingredients, some of which have been enjoyed for centuries, for example, crab, shrimp, oysters, clam juice, white wine, horseradish, and garlic (CharlestonScene 2016; Sellers 2016; Wilson 2017), while others have been enjoyed for only a few decades, like Old Bay seasoning, invented in 1939 in Baltimore, Maryland (Clark 1990). The St. Regis Bar, located in the St. Regis Hotel about two blocks north of the White House, serves a local version of the Bloody Mary called the Capitol Mary (St. Regis 2017). The cocktail uses gin instead of the traditional vodka, because gin is the spirit of choice among the city's social elites. It also incorporates shrimp and signature spices commonly used in preparing a classic Chesapeake Bay crab feast (St. Regis 2018). In North Carolina and South Carolina, where hogs are an important part of regional and local economies, Bloody Mary cocktails incorporate pork garnishes,

FIG. 23. A Bloody Mary and a beer chaser (snit) at a Cafe Hollander restaurant, Milwaukee, Wisconsin. Image courtesy Lowlands Group. Used with permission.

such as barbecued pork ribs, pork rinds, bacon, and ham (Kohatsu 2015; CharlestonScene 2016). Other local garnishes include raw oysters, pickled green beans or carrots, and even a dill pickle spear sprinkled with Old Bay seasoning (CharlestonScene 2016). Bloody Marys in Vermont incorporate local cheeses as garnishes, thus emphasizing the idea of local production through Bloody Mary goût du terroir (Jacobsen 2010, 219; Bartels 2017).

The South and Southwest

There are several iterations of the Bloody Mary in the South and Southwest, particularly in Texas, New Mexico, Colorado, and Arizona, that reflect Latino culture (Savarino 2009; Bartels 2017; Marianella and Fraioli 2017). In these locations it is not uncommon for vodka to be replaced with other liquors or alcohol to achieve a desired taste. For example, a Bloody Mary that contains tequila instead of vodka is referred to as a

Bloody Maria, while a version that uses beer with tomato juice, hot sauce, or salsa is called a Michelada (Filloon 2015; Bartels 2017). Individuals who swap vodka for tequila further celebrate terroir and goût du terroir because tequila makers, much like French winemakers, hail the terroir of the agave plant (Bowen and Zapata 2009). Tequila Ocho, produced in the Mexican state of Jalisco, was one of the first brands to celebrate the diversity of flavor expressed by the agave plant due to its terroir (Moore 2018). Locally grown ingredients such as jalapeños, Hatch green chiles, tomatillos, chipotle peppers, chili powder, limes, and cilantro are used as condiments in regional or local Bloody Mary cocktails (Petro 2014; Bartels 2017; Marianella and Fraioli 2017).

Cajun culture is also well represented through Bloody Mary culinary culture, terroir, and goût du terroir. Bloody Marys in Louisiana, especially southern Louisiana, incorporate some locally produced Creole seasonings: Tabasco pepper sauce, which is produced on Avery Island, Louisiana, and Crystal Hot Sauce, made in Reserve, Louisiana (Bartels 2017). Local garnishes such as okra, sugarcane, and crawfish are also common in Cajun-style Bloody Marys. Sugarcane, an important crop for Louisiana since the plant's introduction there in 1751, is a giant grass that thrives in a warm, moist climate and stores sugar in its stalk (Gravois 2005; Coleman 2013). Before the Civil War most of the sugar grown in the United States came from Louisiana and was harvested by slaves (Coleman 2013). Today sugarcane remains connected with the culture of the region. Louisiana sugarcane is processed into sugar at one of eleven mills located in southern Louisiana (Hilburn 2017). Furthermore, the Louisiana sugarcane industry contributes more than $2 billion to the economy of Louisiana (Gravois 2005). Commander's Palace, a culinary institution and landmark since 1893, is located in the leafy Garden District of New Orleans (Commander's Palace 2018). The Bloody Mary served at Commander's Palace incorporates regional and local items such as Creole seasoning, Tabasco, Crystal Hot Sauce, pickled peppers, pickled okra, and sugarcane skewers (figure 24). The restaurant's recipe for the cocktail is as follows:

COMMANDER'S PALACE BLOODY MARY

1½ ounces vodka

1 teaspoon prepared horseradish

1 teaspoon or two splashes Worcestershire sauce

2 dashes Tabasco and 4 dashes Crystal Hot Sauce

½ cup V-8 or tomato juice

Garnishes:

Creole seasoning (seafood or meat)

1 medium-size pickled pepper, skewered with sugar cane (or cocktail pick)

1 piece pickled okra skewered with sugar cane (or cocktail pick)

Place ice cubes in a tall glass until it's ⅔ full. Add vodka, horseradish, Worcestershire, hot sauce and vegetable juice. Cover the glass with a shaker; shake well, then let rest in the shaker. Wet the rim of the glass and coat the entire rim with Creole seasoning. Pour the drink back into the glass and garnish with the pepper and okra. (Walker 2013)

The Pacific West

In the Pacific West, especially Washington, Oregon, and California, the Bloody Mary culinary culture reflects the connection to the Pacific Ocean and the Asian community (Bartels 2017; Marianella and Fraioli 2017). Bloody Marys in this region often incorporate ingredients such as sake, ginger, sesame seeds, and soy sauce while using garnishes such as California rolls, strips of cucumber, pickled ginger, and wasabi (Filloon 2015; Bartels 2017; Marianella and Fraioli 2017). Farther north, in Alaska, Bloody Marys incorporate food items from local rivers and the Pacific Ocean (Bartels 2017; Marianella and Fraioli 2017). Incorporating seafood garnishes such as sockeye salmon, king crab, and Alaskan oysters expresses to the consumer the local population's close connection to and shared history with the fishing industry (Jacobsen 2010; Bartels 2017).

Conclusion

The purpose of this study was to further elucidate the relationship between production and consumption, as well as identity and place, by examining

FIG. 24. Commander's Palace Bloody Mary, New Orleans, Louisiana. Image courtesy Commander's Palace. Used with permission.

Bloody Mary cocktails. Three objectives guided this research study: documenting the history of the Bloody Mary, examining how Bloody Mary garnishes and condiments have become icons of regional and local culture, and examining regional and local condiments and garnishes for Bloody Marys to highlight variations in culinary culture and goût du terroir.

Since its creation in the 1920s, the Bloody Mary has been enjoyed by tourists, locals, and thirsty, potentially hung-over people everywhere (Bartels 2017). The exact origin of the cocktail is uncertain, and debate remains regarding who created the drink. Fernand Petiot is often credited with creating the modern version of the cocktail in the 1930s around the same time brunch became popular in the United States (Rombauer, Becker, and Becker 2001; Birch 2017). Since the drink's creation almost a century ago, Bloody Mary recipes have incorporated different alcohols and condiments, creating a visually stimulating cocktail.

Bloody Marys are an example of both global and local symbolic fermentation. The cocktail features a frequently internationally produced alcohol, though not all Bloody Mary alcohol ingredients are global, and examples of regional and local alcohols do exist (Olson 2009; Cassel 2018). The garnishes and toppings used in Bloody Marys are often produced regionally or locally and can be used to examine the production and consumption of fermented alcohol products. Production, procurement, and consumption are all linked to culture, history, and tradition of a place (Goody 1982). Locally produced Bloody Marys are embedded in long-standing cultures of consumption where the cocktail's qualities match local notions and perceptions of taste (Flandrin and Montanari 1999; Montanari 2006; Morgan, Marsden, and Murdoch 2009). Furthermore, using local ingredients invokes experiences and memories or expresses a sense of identity (Trubek 2008, 51), and bartenders, using local or regional garnishes and condiments, represent the identity, ceremony, habits, and customs of a regional or local society to locals and tourists for an authentic experience that enables this local knowledge to become nationally and internationally celebrated.

Goût du terroir and culinary culture can be identified in Bloody Marys by examining the local ingredients used in the cocktail. For example, the German, Irish, and Scandinavian immigrants who settled in the North and Midwest regions of the United States brought with them their own cultural traditions and practices. Regional Bloody Marys acknowledge the history of these settlers, especially their breweries, by incorporating local beer chasers, or snits (Olson 2009; Cassel 2018). Garnishes such as venison sticks, summer sausage, bacon, and cheese cubes or curds express and emphasize the regional importance of farming and outdoor lifestyles, especially in Minnesota and Wisconsin (Olson 2009). Additionally, Cajun-style Bloody Mary cocktails that incorporate sugarcane skewers, such as those at Commander's Palace in New Orleans, express not only the economic importance of sugarcane to the Louisiana economy but also how sugarcane is related to Louisiana's history in the time of slavery before the Civil War.

The author has traveled extensively across the United States and enjoyed countless Bloody Marys along the way. This study is an example of the potential avenues available to researchers and scholars incorporating terroir, culinary culture, alcohol, and their production and consumption into local social communities in the United States. Future Bloody Mary studies should consider a more detailed examination of the selection of Bloody Mary garnishes and condiments. Examining the reasons why bartenders choose certain garnishes for Bloody Marys over others, either qualitatively, quantitatively, or both, would allow researchers to document how and why certain ingredients are chosen and how this might impact the overall message or meaning to the consumer buying that cocktail. Finally, future studies should examine the presentation of terroir, goût du terroir, and culinary culture in other alcoholic drinks.

References

Bartels, Brian. 2017. *The Bloody Mary: The Lore and Legend of a Cocktail Classic, with Recipes for Brunch and Beyond.* N.p.: Ten Speed Press.

Birch, Gina. 2017. "The Iconic Bloody Mary." *Times of the Islands* 2 (22): 66.

Bowen, Sarah, and Ana Valenzuela Zapata. 2009. "Geographical Indications, Terroir, and Socioeconomic and Ecological Sustainability: The Case of Tequila." *Journal of Rural Studies* 25 (1): 108–19.

Brunchkateers. 2012. "The History of the Snit." February 4, 2012. https://brunchkateers .com/2012/02/04/the-history-of-the-snit/.

Calder, Ian. 1997. "Hangovers: Not the Ethanol—Perhaps the Methanol." *British Medical Journal* 314 (7073): 2–3.

Cassel, Emily. 2018. "Why Do Bloody Marys Come with a Tiny Beer? We Investigate the Midwest Phenomenon." *CityPages*, January 10, 2018. http://www.citypages .com/restaurants/why-do-bloody-marys-come-with-a-tiny-beer-we-investigate-the -midwest-phenomenon/468498663.

CharlestonScene. 2016. "Lowcountry's Bloody Marys Are Over-the-Top with Garnishes." *Post and Courier*, December 7, 2016. https://www.postandcourier.com /charleston_scene/lowcountry-s-bloody-marys-are-over-the-top-with-garnishes /article_a30a896e-bb1b-11e6-a39e-bb1165792ef0.html.

Chazan, David. 2011. "A Century of Harry's Bar in Paris." *bbc News*, November 25, 2011. http://www.bbc.com/news/world-europe-15887142.

Clark, Kim. 1990. "McCormick Buys Locally Invented Old Bay Crab Spice." *Baltimore Sun*, November 1, 1990. http://articles.baltimoresun.com/1990-11-01/business /1990305022_1_spices-and-seasonings-baltimore-spice-mccormick.

Coleman, Andrew. 2013. "Sugarcane Plantations of Louisiana." Tulane University, April 11, 2013. http://medianola.org/discover/place/987/Sugarcane-Plantations -of-Louisiana (no longer available).

Commander's Palace. 2018. "Our Story." February 2, 2018. https://www .commanderspalace.com/our-story.

Del Casino, Vincent J., Jr. 2015. "Social Geography I: Food." *Progress in Human Geography* 6 (39): 800–808.

Dodge, David. 1942. "Shear the Black Sheep." *Cosmopolitan* 113 (1): 144.

Filloon, Whitney. 2015. "How the Bloody Mary Came to Dominate Brunch." *Eater*, June 13, 2015. https://www.eater.com/2015/6/13/8673499/bloody-mary-explainer -history-variations.

Flandrin, Jean-Louis, and Massimo Montanari, eds. 1999. *Food: A Culinary History*. New York: Columbia University Press.

Goodman, Michael K. 2016. "Food Geographies I: Relational Foodscapes and the Busy-ness of Being More-Than-Food." *Progress in Human Geography* 40 (2): 257–66.

Goody, Jack. 1982. *Cooking, Cuisine, and Class: A Study in a Comparative Sociology*. Cambridge: Cambridge University Press.

Gravois, Kenneth. 2005. "Louisiana's Sugarcane Industry." LSU Ag Center, May 31, 2005. https://www.lsuagcenter.com/portals/communications/publications/agmag /archive/2001/fall/louisianas-sugarcane-industry.

Haigh, Christopher, ed. 1987. *The English Reformation Revisited*. Cambridge: Cambridge University Press.

Henzl, Ann-Elise. 2016. "In Wisconsin, a Bloody Mary Isn't Complete Unless It Has a Beer Chaser." WUWM Radio, September 23, 2016. http://wuwm.com/post /wisconsin-bloody-mary-isnt-complete-unless-it-has-beer-chaser#stream/0.

Hilburn, Greg. 2017. "Sugarcane Where? Louisiana Expands Sweet Spot." *News Star* (LA), September 11, 2017. https://www.thenewsstar.com/story/news/2017/09/11 /sugarcane-where-la-expands-sweet-spot/652915001/.

Houser, Nathan, and Christian Kloesel, eds. 1992. *The Essential Peirce, Volume 1 (1867– 1893)*. Bloomington: Indiana University Press.

IBA (International Bartenders Association). 2017. "How It Started." November 24, 2017. http://iba world.com/howitstarted/.

———. 2019. "Contemporary Classics." Accessed March 21, 2019. https://iba-world .com/contemporary-classics/.

Jacobsen, Rowan. 2010. *American Terroir: Savoring the Flavors of Our Woods, Waters, and Fields*. New York: Bloomsbury.

Jordan, Jennifer. 2008. "Elevating the Humble Dumpling: From Peasant Kitchens to Press Conferences." *Ethnology* 47 (2): 109–21.

Kohatsu, Kimberly. 2015. "26 Crazy Bloody Marys, from A to Z." *Huffington Post*, March 10, 2015. https://www.huffingtonpost.com/Menuism/26-crazy-bloody -marys-fro_b_6841330.html.

LaMorte, Chris. 2016. "3 Bloody Mary Recipes from Chicago's Brunch Scene." *Chicago Tribune*, April 22, 2016. http://www.chicagotribune.com/lifestyles/magazine/ct -mag-mens-2016-drink-print-final-20160420-story.html.

Life Magazine. 1942. "Hollywood Goes Russian." *Life* 13 (8): 35–38.

Marianella, Vincenzo, and James O. Fraioli. 2017. *The New Bloody Mary: More Than 75 Classics, Riffs & Contemporary Recipes for the Modern Bar*. New York: Skyhorse.

Metro-Roland, Michelle M. 2013. "Goulash Nationalism: The Culinary Identity of a Nation." *Journal of Heritage Tourism* 8 (2–3): 172–81.

Montanari, Massimo. 2006. *Food Is Culture*. New York: Columbia University Press.

Moore, Jarrette. 2018. "Tequila Ocho Shares Their Vision of Terroir as 'A Sense of Place.'" United States Bartenders' Guild, May 9, 2018. https://www.usbg.org/browse /blogs/blogviewer?BlogKey=b3cc1858-b7a6-4395-92cc-613643ab7674/.

Morgan, Kevin, Terry Marsden, and Jonathan Murdoch. 2009. *Worlds of Food: Place, Power, and Provenance in the Food Chain*. Oxford: Oxford University Press.

Nelson, Velvet. 2015. "Place Reputation: Representing Houston, Texas, as a Creative Destination through Culinary Culture." *Tourism Geographies* 17 (2): 192–207.

Olson, Drew. 2009. "Ask OMC, Why Do Bloodys Come with Beer Chasers." OnMilwaukee, August 18, 2009. https://onmilwaukee.com/bars/articles/askomcbloodymarys.html.

Park, Michael Y. 2008. "Happy Birthday, Bloody Mary." *The Epicurious Blog*, December 1, 2008. https://www.epicurious.com/archive/blogs/editor/2008/12/happy-birthday.html.

Paulsen, Frank M. 1961. "A Hair of the Dog and Some Other Hangover Cures from Popular Tradition." *Journal of American Folklore* 74 (292): 152–68.

Petro, Brian. 2014. "Classic Cocktails in History: The Bloody Mary." Alcohol Professor, November 13, 2014. https://www.alcoholprofessor.com/blog/2014/11/13/classic-cocktails-in-history-the-bloody-mary/.

Probyn, Elspeth. 2000. *Carnal Appetites: Food Sex Identities*. New York: Routledge.

Rombauer, Irma S., Ethan Becker, and Marion Rombauer Becker. 2001. *Joy of Cooking: All about Breakfast and Brunch*. New York: Scribner.

Savarino, Maggie. 2009. "Search and Distill: Michelada Is Your Standby Beer, Only Better." *Seattle Weekly*, July 14, 2009. http://archive.seattleweekly.com/2009-07-15/food/michelada-is-your-standby-beer-only-better/.

Sellers, Dwaun. 2016. "Where to Get a Good Bloody Mary in Columbia." *The State* (SC), July 13, 2016. http://www.thestate.com/entertainment/local-events/article89204307.html.

Smith, Andrew F. 2007. *The Oxford Companion to American Food and Drink*. New York: Oxford University Press.

Spence, Jordan. 2003. *Complete Bartenders Guide*. London: Carleton.

St. Regis. 2017. "King Cole Bar." Marriott International, November 25, 2017. https://www.marriott.com/hotels/hotel-information/restaurant/details/nycxr-the-st-regis-new-York/6359135/.

———. 2018. "Bloody Mary." February 3, 2018. https://st-regis.marriott.com/culture/bloody-mary-ritual/.

Sutcliffe, Theodora. 2017. "Fernand Petiot." Difford's Guide for Discerning Drinkers, November 23, 2017. https://www.diffordsguide.com/people/51603/bartender/fernand-petiot.

Trubek, Amy. 2008. *The Taste of Place: A Cultural Journey into Terroir*. Berkeley: University of California Press.

Walker, Judy. 2013. "Commander's Palace Bloody Mary Recipe." NOLA Media Group, September 10, 2013. http://www.nola.com/food/index.ssf/2013/09/commanders _palace_bloody_mary.html.

Wilson, Tim. 2017. "Coastal Cocktail Creation . . . Bloody Mary with Steamed Shrimp." *Coastal Carolina Fisherman*, October 11, 2017. http://www.coastalcarolinafisherman .com/2017/10/coastal-cocktail-creation-bloody-mary-steamed-shrimp/.

Farm-to-Bar and Bean-to-Bar Chocolate on Kaua'i and the Big Island, Hawai'i

8

An Industry Profile and Quality Considerations

Ryan E. Galt

Cacao and Chocolate in Hawai'i

Chocolate in Hawai'i is a growing economic sector. While companies like Hawaiian Host have long made chocolate-covered macadamia nuts (from imported cacao and chocolate), cacao is relatively new as a crop being grown commercially in the state, with current efforts dating only from the 1990s.

Previous attempts at growing cacao in Hawai'i commercially did not last long. David Livingston Crawford (1937, 73) notes that a Dr. Hillebrand first introduced cacao to the islands around 1850, with some commercial interest developing in the late 1800s. The Hawai'i Experiment Station planted about three acres of cacao in Hilo in 1905, and in 1917 the state legislature asked the experiment station for an evaluation of cacao as a commercial crop (74). In a prescient passage that predicts important developments today, the experiment station's report to the legislature referenced the uniqueness of the Hawaiian situation in relation to the rest of the world, noting that cacao is grown "on comparatively cheap lands at considerable distances from the world's markets and with the cheap labor so prevalent in most tropical countries. Hawaii, therefore, would have to meet strong competition in marketing the raw product.

By reason of its nearness to markets and its good transportation facilities, Hawaii could counterbalance this disadvantage by entering the field of manufacture—a thing not yet attempted generally by countries far from the consumer" (Hawaii Experiment Station Annual Report 1917, 22, cited in Crawford 1937, 74–75).

A second introduction occurred in the 1980s (Schnell et al. 2005), but the operation fell apart, and in the early 1990s the Hodge Farm planted cacao on one acre in Keauhou, south of Kona on the Big Island. Bob and Pam Cooper bought this property in 1997, added a chocolate-making facility, and started the Original Hawaiian Chocolate Factory. It began producing the first chocolate made in Hawai'i from Hawaiian cacao and was still a successful operation as of 2017 (Bob Cooper, pers. comm., September 1, 2017). The Dole company also started an estate on O'ahu using some germplasm from the failed 1980s operation and produced what Gary Guittard, then-president of Guittard Chocolate, said was "some of the best in the world" (Cheng and Valcourt 2015). That statement is an indication of quality based on the specialty cacao market, in which cacao is "perceived to have a rare or prized flavor profile" (Leissle 2018, 161).

Hawaiian chocolate—cacao growing linked to chocolate making—is therefore a new industry. In 2010 H. C. "Skip" Bittenbender, a cacao and tropical fruit tree specialist at the University of Hawai'i College of Tropical Agriculture and Human Resources, estimated that there were fewer than thirty growers with under one hundred acres of trees (Gomes 2010). Kent Fleming et al. (2009, 9) noted that "the Hawai'i cacao industry is young. All growers are relatively small-scale, and currently none are making a living solely by growing cacao."

Beyond these indications, detailed statistics on the industry do not exist, but it appears to be growing and gathering momentum. The growth today in local Hawaiian chocolate is evident in three forms:

1. Chocolate agritourism is booming, with new tours being added by farms and older tours adding more days of the week to their tour schedules.

Tourists, mostly from temperate regions where cacao does not grow, come to Hawai'i to experience the tropics and are increasingly interested in where food comes from.

2. The number of people making chocolate in Hawai'i is growing rapidly, with many new makers getting into the business and more variety available in local grocery stores. Outside of the islands, producers and consumers in the craft chocolate world know Hawai'i-grown chocolate because of two well-known craft chocolate makers, Mānoa and Madre, both on O'ahu, the most populous and most visited island. Neither use exclusively Hawai'i-grown cacao due to its scarcity, but both regularly feature excellent bars made from high-quality specialty cacao beans from the Big Island.[1] Both have won numerous national and international awards, putting Hawai'i-made and grown chocolate on the world map of fine chocolate. In addition to these two well-known makers, there are more than twenty others, most producing on a much smaller scale.

3. More land is being dedicated to growing cacao, even though the total amount is still very small (Gomes 2010). Many of the new makers are also cacao growers who make single-estate chocolate, a relative rarity in the world, because since colonial times there has been a vast geographical disconnect between cacao growing and chocolate making (Coe and Coe 2013). But there are many other cacao growers who have recently started operations in Hawai'i, from commercially oriented estates to householders growing cacao in their backyards. Cacao landscapes, however, are mostly not readily visible, since cacao trees are an understory species (they evolved under the canopy of the western Amazon rainforest) and like to be protected from wind, unlike grapes in the landscapes of wine-growing regions (Stanislawski 1970).

I explore these transformations in this chapter by focusing on farm-to-bar and bean-to-bar chocolate makers on Kaua'i and the Big Island.[2] In reporting on my findings, I engage in a rather unorthodox mixing of my data and findings, using traditional forms of knowledge production in social science as well as personal and panel-tasting experience. The chocolate makers I studied are much smaller in scale than Mānoa and

Madre, with many starting recently and others just getting going. In this chapter I argue that promising developments for cacao growing and chocolate making abound but that a substantial minority of chocolate makers face quality problems. By craft chocolate standards, the products of Hawaiian chocolate makers range from very high to very low quality. The conclusion explores questions around quality and standards and issues involved in applying them.

A Primer: The Fermented Landscapes of Specialty Cacao and Craft Chocolate

While most people know that alcohol production requires fermentation, few know that chocolate is a food much improved by fermentation. The flavor of "finished chocolate comes from a multitude of sources: the genetics of the trees; terroir, including the land and soil conditions as well as climate during pod maturation (5½ months long); fermentation and drying; the local microbiome where the fermentation is being carried out; and the factory processes such as roasting, conching, etc." (Seguine and Meinhardt 2014, 25). While fermentation is not required to turn cacao beans into chocolate, it is "a crucial step for flavor development," and the industry consensus is that fermented beans are required for high-quality chocolate (Leissle 2018, 6). Indeed chocolate industry professionals argue that fermentation is a much more complex and important process for fine chocolate than it is for fine wine (Smillie 2016). To understand why, some background information is important.

Chocolate is made of finely ground dried cacao "beans," which are the seeds within the cacao tree's fruit, the cacao pods (fig. 25). Cacao trees typically begin to yield pods from three to five years after planting. Each cacao pod contains thirty to fifty cacao beans, in a matrix of white, sweet, mucilaginous pulp, which contains the sugar used as the fuel for fermentation (fig. 26).

Fermentation for cacao starts when the pods are broken open and the contents are placed in a pile or box. Fermentation style and quality vary a great deal between types of markets. The largest distinction in

FIG. 25. Cacao pods, the fruit of the cacao tree, hang on the trunk and main branches in a cacao grove north of Honomū, Big Island. Photo by Ryan E. Galt.

FIG. 26. A cacao pod cut in half to reveal the cacao "beans" (seeds) in their sweet, white pulp, photographed north of Hilo, Big Island. The beans hang on the placenta, strands of which are visible at the very top. Photo by Ryan E Galt.

cacao markets is between "bulk" cacao and "specialty" cacao (also known as fine flavor cacao). Bulk cacao forms the backbone of the chocolate industry and is mostly grown in West Africa (Leissle 2018). It is priced through the cocoa commodity price set by the New York and London Stock Exchanges. No one knows exactly what proportion of bulk cacao is fermented because it often matters little to the final product. When bulk cacao is fermented, the process is not usually tended as attentively as for specialty cacao because there is almost no incentive for high quality (O'Dougherty 2017) except in Ghana, known for its high-quality bulk cacao. Thus some bulk cacao is unfermented, or "washed," with the pods broken open and the pulp washed off the beans soon after harvest and therefore not subjected to fermentation in heaps or boxes (Leissle 2018). Specialty cacao, on the other hand, has various flavor qualities that distinguish it from bulk cacao. The International Cocoa Organization (ICCO) considers the following tasting notes when differentiating beans for placement in the "flavor" category: "fruit (fresh and browned, mature fruits), floral, herbal, and wood notes, nut and caramelic notes as well as rich and balanced chocolate bases" (ICCO 2019). Fermentation, genetics, and terroir all make flavor contributions to specialty cacao (Sukha et al. 2014). Chocolate makers are willing to pay up to five times the bulk cacao price for this specialty resource, from about $3,000 to $10,000 per ton (Nieburg 2016).

Without considerable manipulation, unfermented (washed) cacao beans make disgusting, unpalatable chocolate. This is not an observation made in the scientific literature on chocolate flavor, as that literature has been focused on producing palatable chocolate flavors out of bulk beans of varied quality (fermented and unfermented). I heard this sort of comment from many craft chocolate makers during interviews (see Galt et al. in preparation) and had to take their word on this quality distinction until I experienced it myself. At a workshop demonstrating the importance of fermentation during the 2016 Northwest Chocolate Festival, chocolate makers Dandelion and Raaka made chocolate from unfermented beans, emphasizing that this was for education purposes only and that they would never make chocolate that tasted like this nor let their customers eat it. To me, it tasted like a strong mix of vomit and tar, with hints of gasoline. It was so unappealing that it worked as an appetite suppressant for me, mimicking the experience of having recently vomited and thus making me uninterested in eating for about two hours. Many craft chocolate makers explain that the large industrial chocolate makers like Hershey's have spent considerable time, money, and effort to transform unpalatable raw materials—unfermented and poorly fermented bulk cacao—into a mass-produced, standard product. Food science has taught industrial makers to highly process cacao (e.g., roast it at high temperatures to destroy off-flavors and add many extra ingredients, like sweeteners, milk powders, and emulsifiers). In talking about this process, Colin Gasko, head of the well-respected craft chocolate maker Rogue Chocolatier, noted, "I don't know how they're able to make a palatable product out of such awful raw material. It's some impressive chemistry" (quoted in Jacobsen 2010, 260).

In contrast, craft chocolate makers concentrate on sourcing specialty cacao, which is always fermented. Thus, cacao farms that grow specialty cacao are part of fermented landscapes (see chapter 1, this volume). Farmers connected to these markets must pay attention to quality, as "several characteristics of chocolate strongly depend on the processes done at the very beginning of the supply chain" (Saltini, Akkerman, and Frosch 2013, 168). Cacao fermentation is arguably the most crucial process, as it creates flavor precursors for the sought-after flavors and reduces bitterness (Afoakwa et al. 2008). Cacao fermentation generally requires relatively large volumes—at least half a cubic meter or yard—to reliably reach high enough temperatures to chemically transform the beans (O'Dougherty 2017). It is done in piles, in pits, and, perhaps most commonly in the specialty cacao world, in wooden fermentation boxes (fig. 27). These are typically about one meter by one meter and made of wood slats to allow for air exchange. In well-tended ferments the beans are turned about every day to enhance aeration (thereby allowing the oxygen-loving microorganisms to remain dominant) and left to ferment from three to seven days, depending on the cacao variety, until the transformation is deemed complete. Fermenting for too long, or for not enough time, "is also noticeable—and not deliciously so—in the chocolate's flavor" (Leissle 2018, 171). Cacao bean fermentation is an art that, when done successfully, transforms the bitter raw cacao bean into a flavorful bean ready to be made into chocolate or roasted and sold as snacking cacao or nibs, which are roasted beans that have been cracked and winnowed.

Once fermented or washed, cacao beans are then dried. Drying processes vary greatly, from spreading the beans on the side of the road to placing them on wooden drying racks (fig. 28). It is in this form—dried beans—that cacao is usually sold from growing regions to making regions, thus providing the raw material for chocolate makers.

Dried cacao beans are heavily transformed in the chocolate-making process. Through careful processing—including sorting, roasting, cracking, winnowing, grinding/refining/melanging (and sometimes conching), and tempering—the flavor notes produced and enhanced through fermentation can be further showcased. Refining, also called grinding and melanging, is the process of reducing the particle sizes of cacao solids (and sugar, if added) to particles small enough to be felt as smooth by the human tongue. Conching, "probably the least understood process in modern chocolate making," is the process of heating and mixing all of the chocolate ingredients once combined, with the goal of mellowing the sharp taste of fresh cacao and reducing acidity and bitterness (Nanci n.d.). If undertaken by experienced hands, these processes allow the eater to distinguish a variety of flavor notes in one savored bite.

While most of the chocolate-consuming population has yet to explore many of the flavor nuances of craft chocolate, that time is coming. As with the rise of "craft" production and consumption for wine, beer, coffee, and cheese in recent decades, craft chocolate is having a moment. Craft chocolate bars, ranging in price from about four to eighteen dollars, are increasingly appearing on supermarket shelves in the United States. While there were only a handful of craft, bean-to-bar chocolate makers

FIG. 27. Wooden cacao fermentation bins, considered a best practice for fermentation, photographed south of Kona, Big Island. Photo by Ryan E. Galt.

FIG. 28. Cacao beans drying on wooden and mesh racks, photographed south of Kona, Big Island. Photo by Ryan E. Galt.

in the country in 2006, there are now close to three hundred (Hoehn-Weiss 2018). In the 2010s these makers have focused on highlighting the various flavor qualities of cacao beans from different countries and regions (Williams and Eber 2012; Leissle 2013a, 2013b; Giller 2017). This makes it possible to appreciate the strong citrusy and/or raspberry notes of a Madagascar bar or the fruit-forward berry jam character of a Dominican Republic bar, giving rise to the idea of terroir, or taste of place (Trubek 2008) within chocolate (Nesto 2010; Galt et al. in preparation). And this is just the beginning of the distinctions one can taste.

These developments made me curious about the connections between craft chocolate and cacao growing in Hawai'i, the only U.S. state where it can be grown, given the crop's tropical climate requirements. Since European colonial times, cacao-growing landscapes have been very far removed from the chocolate-making process, since almost all chocolate making is done in the Global North (Leissle 2018).[3] Thus, the typical connections between growing and consuming are much more elongated than for other commodities, especially alcohols, considered in this volume. Yet, as Kristy Leissle (2018, 44) notes, "Hawaii—the only place [state] in the US where cocoa grows—is making an innovative contribution, as small-scale manufacturers condense the commodity chain by making chocolate from local beans." This makes it an exciting location in which to study chocolate.

Craft Chocolate Quality Norms

The original goals of my research in Hawai'i included gaining a broad understanding of the possibilities and challenges facing the farm-to-bar and bean-to-bar movement within the relatively remote contexts of Kaua'i and the Big Island. Before starting, I considered the tasting experience to be a pleasant addition to the research, rather than a central issue to be explored. This changed after a few days of fieldwork, since the range in quality of the chocolate I ate was extreme from the beginning, and this wide range—from excellent to very bad—continued throughout my experience. Since the ideas of "excellent" and "bad" are highly subjective, below I lay out a framework for how chocolate quality has been defined in the recent craft chocolate movement. This is the framework for understanding my comments on quality in this chapter.

Chocolate quality is notoriously difficult to define. The craft chocolate industry has yet to agree upon industry-wide quality standards at any level, both for raw materials—wet beans and fermented beans—and for finished chocolate bars—temper, off-flavors, and so on.[4] This apparent hesitancy in establishing standards has occurred in large part because the craft side of the industry celebrates and accentuates the radically different flavors produced by cacao, one of the world's most complex foods. Craft chocolate has defined itself largely in opposition to mass-produced chocolate, where consistency in flavor profile across time has been a goal. However, in place of formalized norms are norms of the informal, unwritten sort. After having relentlessly tasted craft chocolate for the last four years, and having tasted quite a bit of chocolate in Hawaiʻi that violated these norms, I have experienced the unwritten norms of the craft chocolate movement and describe them below.

The flavor norms for U.S. craft chocolate are strongly in line with the ICCO definition of specialty cacao noted above. These norms include:

- a strong preference for flavors that work well with sweetness according to the U.S. palate, especially fruity, nutty, and "chocolatey," which is a craft industry term to describe the "standard" chocolate flavor that U.S. consumers are used to and that is produced in its highest-quality form in Ghana (Leissle 2018);

- a mixed take on flavors not historically associated with chocolate by the U.S. palate, such as woody, earthy, floral, leather, and mushroomy (these can be found as the main flavor notes in some bars, but they are relatively rare);[5]

- aversion to the dominance of flavors that almost all would agree are stomach-churning "off-flavors" for chocolate, such as vomit, gasoline, tar, vinegar, and so on, which are considered to be errors in the making process, rather than something that might be tolerated, as in wine (see, for example, the Good Food Awards chocolate tasting rubric in table 6); and

- complex and varied flavor notes (for the best bars), which come in and out of the tasting experience over time, like sections of an orchestra playing a symphony. Connoisseurs and leading chocolate makers and experts in the industry, including Ed Seguine, Georg Bernardini, Art Pollard, and many others (Bernardini 2015; Ed Seguine, pers. comm., 31 May 2016), highlight this "symphonic" quality of the best bars, and it is used in chocolate-tasting competitions such as the Good Food Awards (see table 6 under Methods).

Other craft chocolate norms include:

- a strong temper with a good snap and a surface sheen (the goal, technically speaking, is fat crystal form V), with fat blooming and cocoa butter rancidity considered to be flaws;
- high-quality packaging with an in-depth product description often going into flavor profile, the beans' place of origin, the grower/cooperatives involved, and so on; and
- experimentation and innovation, with the goal of making chocolate in a way unlike that used for mass-produced chocolate and with a focus on highlighting terroir, or a taste of place (Jacobsen 2010; Nesto 2010; Leissle 2018) and, thus far, a willingness to share and collaborate as small chocolate makers.[6]

This is the quality framework that I applied to my tasting experiences with farm-to-bar and bean-to-bar chocolate in Hawai'i. Most of the tasting panelists, detailed below, are familiar with it.

While the quality norms of U.S. craft chocolate seem like an external imposition, I think they establish an appropriate standard for a few reasons. First, Hawai'i is part of the United States, and the United States is ground zero for the craft chocolate movement, with innovation and quality now surpassing that of chocolate made in Europe, according to Georg Bernardini (2015). Second, priced at six to fifteen dollars per bar, farm-to-bar and bean-to-bar chocolate in Hawai'i fits within the price range of mainland craft chocolate, which is between four and twelve dollars (with rare extremes going to eighteen dollars). This puts

Hawaiian chocolate squarely within the sales niche of craft chocolate rather than mass-marketed chocolate bars like Hershey's. Third, Hawaiian cacao beans are typically considered specialty beans, and they have won multiple international flavor awards recently. Two of the top eighteen winners of the 2017 Cocoa of Excellence program's International Cocoa Awards—a widely recognized industry award given at the premier international chocolate venue, the Salon du Chocolat in Paris—were from the state (from the University of Hawai'i, Mānoa, and the Nine Fine Mynahs Estates on O'ahu), and another of the top fifty was from Steelgrass Farm on Kaua'i (Ernst 2017). Hawai'i cacao expert Skip Bittenbender notes that "Hawai'i grown cacao produces a high quality chocolate with a superior flavor profile" (quoted in Ernst 2017).

Methods

In August and September 2017 I conducted fieldwork on Kaua'i and the Big Island. I chose two groups of chocolate makers on the two islands: (1) farm-to-bar makers (who grow their own cacao and make it into chocolate), and (2) Hawai'i-focused bean-to-bar makers (who make chocolate using only cacao grown in Hawai'i). Based on internet searches, snowball sampling, and local guides, I was able to find six of these makers on Kaua'i and seventeen on the Big Island. I conducted interviews with a majority of the members of these two populations—three on Kaua'i and fourteen on the Big Island. The interviews focused on their business model, their views of the craft chocolate movement, cacao sourcing, certifications, competition, sales, consumers, finances, labor, and challenges. Interviews typically lasted about an hour. I also visited many cacao farms and chocolate factories, workshops, and kitchens to make observations.

I also gathered chocolate to eat and then engaged in personal tastings, an experience that informs my findings. The individual subjective aspect of this tasting is undeniable, as I have my own taste preferences, which have been largely shaped by enjoying mass-produced chocolate as a child and into adulthood and then by relentlessly tasting craft chocolate— easily eating more than one hundred bars per year—over the last four

TABLE 6. Good Food Awards chocolate judging rubric

	5	4	3	2	1
Aroma (15%)	Enticing, even irresistible. Compels tasting. Complex but balanced (not overpowering). Aromas are natural, not artificial.	Convincing aroma. Clear and attractive. Somewhat complex.	Not unpleasant but lacking in complexity. May be weaker or stronger than it should be. Neither compels tasting nor discourages tasting.	May not have a distinguishable aroma. May have a somewhat challenging aroma.	Bad odor. Disagreeable. May be alarming. Reveals defects.
Flavor (45%)	Takes the taster through a wonderful arc or progression of flavors. Exhibits subtlety, delicacy, and sophistication. Displays balance of flavor components. Lingering, desirable finish. Compels contemplation.	Fairly sophisticated arc or progression of flavors. Some flavors may be out of balance, but the experience of tasting remains very enjoyable and interesting.	May be flat, monotone, or otherwise unremarkable. Flavors may be out of balance, but the experience of tasting remains acceptable.	Unpleasant qualities outweigh pleasant qualities. May have a variety of flavor notes, but some are unbalanced, unpleasant, or unremarkable. Intensity may be too high or too low.	Bad flavor, undesirable in chocolate. Disconcerting. Reveals clear defects. Leaves unpleasant aftertaste.

Column header row (top): Excellent. Would recommend highly. | Very good. Would recommend. | Good. Adequate but unexceptional. | Flawed. Would not recommend. | Seriously flawed. Would discourage.

Mouthfeel (20%)	Enticing, even irresistible. Compels contemplation. Perfect complement to the flavor. Exemplary melt.	Complements the flavor well. Melts nicely. Adds to the enjoyment of tasting.	Neither enhances or detracts from the overall experience. Not much special to recommend it.	May be challenging or unpleasant. May be sticky, muddy, chalky, or waxy.	Bad mouthfeel, undesirable in chocolate. Reveals serious defects. Leaves disagreeable feeling.
Overall impression (20%)	Masterful. Reveals exquisite craftsmanship. Unequivocally deserving of an award.	Very strong example of fine chocolate but not exceptional.	Acceptable but not very good.	Limited appeal. Needs improvement. Negative elements outweigh the positive.	Unacceptable.

Source: 2017 Good Food Awards, https://goodfoodfdn.org/awards/winners/2017/.

TABLE 7. Hawaiian craft chocolate data sources

	INTERVIEWS	FARM VISITS	FACTORY, WORKSHOP, AND KITCHEN VISITS	PERSONAL TASTINGS (AUTHOR)	PANEL TASTING (TASTER $N = 9$)
Farm-to-bar makers	13	12	3	13	12 (15 bars)
Bean-to-bar makers	4	—	1	2	2 (2 bars)
Total	17	12	4	15	14 (17 bars)

years. Thus, in addition to recording my tasting impressions—based on quality as defined in the craft chocolate movement—I brought many bars home and set up a panel tasting with nine of my graduate students. These students tasted seventeen bars of farm-to-bar and bean-to-bar chocolate from Kauaʻi and the Big Island. Most of them have eaten considerable amounts of craft chocolate, since I bring in craft chocolate bars made in the United States and elsewhere to share in our weekly lab group meetings. Thus, my general tasting palate and that of my students have been informed by the standards of the craft chocolate movement but in an informal way, except for two students who have been chocolate judges in the Good Food Awards (one of whom has also been active in helping to create sensory standards for chocolate tasting). The panel used the tasting protocol from the Good Food Awards chocolate judging rubric (table 6), except that the packaging was available for them to see. The panelists individually recorded their sensory impressions of each bar—quantitative and qualitative— within their own tab in a Google spreadsheet set up for the tasting.

Table 7 shows the number of interviews, visits, and tastings as data sources that inform this chapter. I analyzed the interviews and observations according to patterns observed. The panel tasting data were compiled and analyzed according to the composite scoring done in the Good

Food Awards. I present my findings in two sections below: (1) context, value-chain models, and agritourism and (2) quality considerations.

Context, Value-Chain Models, and Agritourism

A Chocolate World of Its Own

Hawai'i is the most remote island chain in the world yet is also part of the United States. Given the history of Hawai'i as a kingdom, a colony, and a U.S. state, it is important to note the identities of those involved in cacao and chocolate. Only one of my seventeen farm-to-bar and bean-to-bar interviewees identified as having Native Hawaiian ancestry. The vast majority were of European descent (commonly called "haoles," which also refers to foreigners and foreign things generally), yet it is common among white farmers to use Hawaiian words, including *kokoleka*, in their business or product names. That cacao is a recent introduction and not a "canoe plant"—brought by the first Polynesian settlers of the islands—makes it unsurprising that it does not have a large role within the renaissance of Native Hawaiian agriculture. Whether Native Hawaiian farmers embrace it as a crop in the future will be interesting to see.

Hawai'i is the only tropical state in the nation (the United States has a number of tropical territories, but no others are states), and it has an economy largely dependent on tourist spending. These characteristics mean that its cacao and chocolate production is a world unto itself in many respects, especially in labor availability and land prices.

Unique aspects emerged when I asked makers about the major challenges they face. Farm-to-bar makers commonly brought up the high cost of farm labor, as well as an overall shortage of qualified workers (see also Fleming, Smith, and Bittenbender 2009, 4). Generally, the small scale of production for smaller grower-makers is such that they and their households provide most of the growing labor, including planting, pruning, fertilizing, and harvesting. The harvesting and postharvest processing—cracking the pods, scooping out the seeds, and getting the seeds into fermentation containers—tends to require the most labor in a time-sensitive window, as allowing the pods to sit too long after harvest can negatively affect flavor. Work groups are a common solution to labor shortages at harvest. Having given up on finding regular employees, one grower-maker creates occasional work groups of family and friends, guaranteeing that all will be paid twenty dollars per hour. Another maker relies on work parties that are formed around learning about cacao and chocolate making.

An additional challenge is that all agriculture in Hawai'i faces high land prices, as tourism and the desire to live in a tropical climate in the United States have driven real estate prices higher. High labor and land prices mean that producing cacao in Hawai'i is very expensive relative to the rest of the world. One economic analysis shows that costs of production per acre are $7,927 (Fleming, Smith, and Bittenbender 2009, 11), which is equivalent to $4,560 per ton of dried/fermented beans, although some growers have estimated costs as being nearer to $10,000 per ton (Klassen 2016). These costs are about two to five times above the bulk chocolate commodity price, which tends to be around $2,000 to $3,000 per ton.

Thus, the Hawaiian context creates certain economic conditions that shape chocolate production. For bean-to-bar makers cacao as a raw material comes at a much higher cost than bulk cacao world market prices, which means that they must also sell their chocolate at craft chocolate prices. Farm-to-bar makers, who attempt to make a living as farmers in Hawai'i, need to use their farms in a way that generates considerable income. The two major strategies of farm-to-bar makers are to (1) engage in value-added production through chocolate making, and (2) create agritourism opportunities or find another way to connect to the tourist economy. While these attributes are similar across makers, I found a wide variety of models in the way that Hawai'i-grown cacao beans get into consumers' mouths as chocolate. I explore these below.

Diverse Value-Chain Models
An important finding of this research is the diversity of forms that the value chain of farm-to-bar and bean-to-bar chocolate takes in Hawai'i.

This diversity manifests across all sections of the cacao-chocolate value chain: cacao growing, cacao sourcing, chocolate making, and chocolate sales/exchange.

Cacao orchards on the two islands in question run almost the full range of the shade continuum identified by Robert Rice and Russell Greenberg (2000). There are very diverse polycultures based on cacao having been interplanted with a variety of other fruiting trees and shade trees (fig. 29), monocultures of cacao planted under sparse shade (fig. 30), and full-sun monocultures (fig. 31). Growers have very different opinions about the use of shade. Those growing full-sun monocultures highlighted that they were at the northern limit of the global cacao production zone, so the trees need less shade than in regions closer to the equator. Those engaging in shaded polycultures highlighted an organic and agroecological approach to farming, emphasizing local nutrient cycling rather than synthetic fertilizers, as well as the origins of the cacao tree as an understory plant. Land tenure was similarly diverse and ranged from those owning their land (for various time periods, including multigenerational holdings and as part of the back-to-the-land movement in the 1970s), to temporarily growing on land that might be developed for condos, to convincing other farmers to plant some cacao groves in their already diversified fruit forest.

Sourcing of cacao is highly varied as well. The single-estate model of sourcing involves makers who use only cacao beans produced on their farms. A number of farm-to-bar makers rely on this model and are able to control quality from farm to bar. For makers who have a good handle on producing quality cacao and chocolate, this approach yields exceptional bars. However, for those less practiced or taking less care in the processes involved, especially fermentation, the single-estate bars highlight their limitations, as noted below.

Another sourcing model is a distributed farming model, whereby farm-to-bar makers have a known going rate (ranging from one to two dollars) for purchasing pods from growers and encourage householders and small-scale farmers to plant cacao to feed into their chocolate-

FIG. 29. A polyculture: this young cacao grove was planted among shade trees and fruiting trees, photographed in northeastern Kauaʻi. Photo by Ryan E. Galt.

FIG. 30. A cacao grove planted under a light shade layer in northeastern Kauaʻi. Photo by Ryan E. Galt.

FIG. 31. A full-sun monocultural cacao grove, photographed in eastern Kauaʻi. Photo by Ryan E. Galt.

making enterprise. The quality achieved through this model varies greatly, as with the single-estate model. Farm-to-bar makers who know a great deal about growing cacao and fermentation provide technical assistance to the backyard growers who sell them pods and then ferment in a central location to achieve volume. One maker in particular produces exceptional bars from specific groves with unique genetic traits (referred to on the islands as "Criollo" but without genetic testing to back up the claim that it belongs to this particular grouping [see Motamayor et al. 2008]). These bars are not available through retail channels because they are extremely rare and cost twenty to fifty dollars each and therefore are available only by word-of-mouth connections. On the other hand, another maker paying less attention to quality produces off-flavored chocolate from such a distributed model. This might result from not providing assistance to growers and/or a lack of concern for ripeness and/or the time between harvest and fermentation (waiting too long can reduce cacao quality); however, issues of fermentation quality abound as well, as noted below.

Chocolate-making factories and kitchens are similarly diverse: some are medium-scale operations (fig. 32), but most are small. With rapid advances in small-scale chocolate-making equipment and knowledge, and with the knowledge shared by John Nanci on his *Chocolate Alchemy* blog and website, setting up a maker-ready chocolate factory with making equipment can be done for as little as $1,000 to $2,000. Most bean-to-bar and farm-to-bar makers followed this model and established a small chocolate-making facility in their home kitchens. A less common practice, done by two single-estate makers, is to grow the beans and then contract the chocolate making out to world-class chocolate makers on Oʻahu, shipping their beans there and receiving finished, wrapped bars in return. This practice produces exceptionally good chocolate.

Chocolate makers in Hawaiʻi face an important tropical challenge: a lack of climate control makes tempering difficult. Tempering chocolate is an essential part of the making process. Cocoa fat crystals can take six specific forms, and the goal of most chocolate making is to achieve

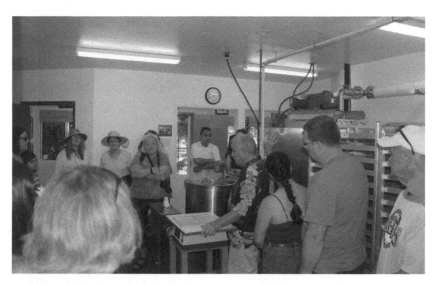

FIG. 32. Bob Cooper giving a tour at the Original Hawaiian Chocolate Factory, Kona, Big Island. Photo by Ryan E. Galt.

Form V (also known as beta crystals), which gives the surface a sheen and provides enough rigidity for the chocolate bar to "snap" when broken into pieces (Notman 2015). Given the year-round warm temperatures on the islands, moderated by the trade winds, residences rarely have air-conditioning. Cooler temperatures, created easily by air-conditioning, make the tempering process much easier in a tropical climate. Lacking air-conditioning, many chocolate makers reported waking up in the middle of the night and very early in the morning to temper in their kitchens during this coolest period of the day. This practice yielded highly variable results, from good tempers to poor tempers with a crumbly texture, which shows that a consistent fat crystal Form V has not been achieved.

Another way to facilitate tempering is through creating climate-controlled shared kitchen space. While shared commercial kitchens have been increasing in number as the craft food movement expands (including in Hilo, the Big Island), chocolate makers tend to avoid these spaces. The chocolate-making process requires very long periods for some processes (e.g., running a melanger for twenty-four to seventy-two hours), and chocolate easily takes on the odors of other foods (e.g., if someone is cooking garlic in the shared space while the melangers are running). One interviewee on the Big Island was establishing a commercial kitchen with climate control to be shared specifically by chocolate makers for a simple production fee. Such a space could be a large boon to makers lacking their own space to temper their chocolate well.

Makers use a variety of retail venues to sell their chocolate. Getting into retail outlets—mostly the islands' few grocery stores, especially the natural food stores and gourmet stores—is a common strategy and seems to serve both the makers and the grocery stores well. Grocery stores tended to have sections of a variety of locally made tropical items (fig. 33). Indeed in almost all grocery stores one can find some Hawai'i-made farm-to-bar or bean-to-bar chocolate, although these invariably receive less shelf space than the imported chocolate from mainland companies (fig. 34). In other stores Hawaiian chocolate means the mass-produced, chocolate-covered macadamia nuts from Hawaiian Host or other such confections using imported couverture or imported cacao beans.

Direct-to-consumer sales are also a common market outlet because of agritourism. Makers running farm tours always end the tours in their gift shops, farm stands, or lanais (roofed, open-sided porches). One maker who offers farm tours also runs a roadside farm stand—with chocolate and various chocolate-related gifts like hats and shirts—on the main road on Kaua'i, thereby receiving tourists looking for chocolate. Additionally, some makers sell in farmers' markets. Another maker sells through his coffee shops on the Big Island (fig. 35).

Chocolate also is exchanged in a less commoditized form. One farm-to-bar maker uses his chocolate as a special treat for guests who come to his retreat center and as a part of the chocolate dance/cacao harvest parties thrown each Friday night. Another who runs an upscale Airbnb is considering giving it as a gift for her guests. A larger farm-to-bar maker gives chocolate away during his fifty-dollar chocolate-making classes, which start with a harvest and go all the way through the processing

FIG. 33. Locally made chocolate on display at Healthy Hut, a health food store, in Kilauea, Kauaʻi. Photo by Ryan E. Galt.

FIG. 34. Mainland and imported chocolate on display at Foodland, a chain grocery store, in Princeville, Kauaʻi. Photo by Ryan E. Galt.

FIG. 35. One grower of coffee and cacao has a small chain of cafés on the Big Island (including Hilo, pictured), where his coffee and chocolate are available. Photo by Ryan E. Galt.

of the cacao. All of these exchange relationships—both commodity exchange and less commoditized exchange—reflect the strong connection between chocolate and tourism in Hawai'i.

Chocolate Agritourism

The state's economy depends in large part on tourism, with millions of visitors arriving each year. Chocolate in Hawai'i is strongly connected to tourism; while high land and labor values are related outcomes, tourism also brings in consumers for cacao farm tours and chocolate. Indeed, when asked about the level of demand for their chocolate, many makers said that they were always able to sell everything that they make. The novelty of a U.S.-grown-and-made chocolate bar proves irresistible to many mainland U.S. consumers.

In addition to attracting tourists who are eager chocolate consumers, cacao farm and chocolate tours offer a new experience to visitors from the mainland or from other temperate climates. Steelgrass Farm and Garden Island Chocolate on Kaua'i and the Original Hawaiian Chocolate Factory and Kuaiwi Farm on the Big Island all offer tours of various lengths, easily found online via major tourism websites like TripAdvisor and Yelp. Other grower-makers are creating their own tours as well after seeing both the economic opportunity of garnering tourist dollars and expanding their customer base to longer-term consumers who order products from the farm once they return home.

I took three cacao farm tours aimed at tourists. Two were similar, three-hour tours for seventy-five dollars each.[7] They each began with a leisurely stroll through highly biodiverse cacao farms, and the tours highlighted both cacao and other tropical crops, especially fruit trees and vanilla. They both involved a wide-ranging tasting of tropical fruits and nuts and ended with a chocolate tasting of a variety of products. The third tour was shorter, one hour, and focused on the cacao trees, the drying racks, the fermentation bins, and the chocolate factory itself. Tastings were included as part of the tour, although the formats varied. Two of the tours provided samples of only their chocolate, while the other one provided samples of not only their chocolate but many other makers' products in order to contextualize their chocolate in the realm of craft chocolate.

These tours are a valuable educational opportunity for chocolate eaters since, as a commodity, chocolate has been subjected to an extreme form of commodity fetishism (Martin and Sampeck 2015). First, while some food commodities come from production processes conducted nearby (e.g., oranges grown and eaten in California), cacao grows commercially only in the tropics, while the vast majority of consumption occurs in the temperate regions. This means that most consumers have not seen a cacao tree let alone know what cacao is and how it as a raw input relates to the final chocolate product. Second, until recently almost all cacao has been highly processed before being added to a food and has been a lesser ingredient in products composed largely of other things (e.g., a Hershey's bar has only 11 percent cacao content, being composed mostly of sugars and milk powder [Sethi 2015]). This means that the material in its unprocessed form created by the tree—the cacao "bean" (a seed, botanically)—has also not been personally observed by most of its consumers.

Tourists taking the tours were extremely eager to learn more about tropical crops, especially cacao, since the plant was so foreign to most of them. Starting from almost no knowledge base, the visitors had experiences that quickly taught them and helped dispel some of chocolate's commodity fetishism by showing the biological origins of cacao. Additionally, on two of the three tours I attended, some parts of the labor process were discussed, with growers highlighting some of the challenges. Yet this discussion occurred in a sanitized form, with labor exploitation, if acknowledged, presented as something happening in other cacao-growing regions.

All of the chocolate tours offered a major leap forward for consumers' understanding of cacao as a tropical fruiting tree and of the different flavors chocolate may have. Whether this creates a positive image of all chocolate, or just Hawaiian chocolate, would be an interesting question to investigate with further research. At the same time, chocolate agritourism greatly benefits those operations that rely on it as an income stream. Trent, one of the farm-to-bar makers operating tours, stressed the importance of agritourism: "Tourism holds an important key to making farming profitable again in Hawai'i . . . it's taking the visitor dollar and giving it to the farms, right? Not only for value added, but the visitor dollar is huge in Hawai'i. It's almost $2 billion a year spent on this island [Kaua'i]. So if we want to revitalize agriculture, it's finding an easy way to connect people, which they love and the farmers love."[8] Additionally, agritourism creates a direct connection and possible feedback loop between chocolate makers and chocolate eaters, which certainly enhances quality, discussed in the next section.

Quality Considerations

According to Michael Laiskonis, creative director for the Institute of Culinary Education, "Making chocolate is easy. Making good chocolate is extremely difficult" (Giller 2017, vi). Because of factors uniquely present in Hawai'i—ready access to capital and chocolate-making equipment, high-quality transportation infrastructure, access to a land-grant university with a cacao program, the close proximity of growing and making, and considerable consumer demand—and the state's reputation for specialty cacao (Ernst 2017), I was expecting my fieldwork to lead me to some of the best chocolate I've ever eaten. And indeed I did eat what I consider to be some of the world's best chocolate, on par with what chocolate connoisseurs like Georg Bernardini (2015) consider to be the best.[9]

What I was surprised by was that I also ate bad chocolate, the worst chocolate I've ever eaten. My use of the concept of "bad chocolate" is not chocolate snobbery per se, in that chocolate snobbery might equate Hershey's-style chocolate with bad chocolate, which is not what I intend here. Rather, what bad chocolate means here is that it deviates strongly from the unwritten norms of the U.S. craft chocolate industry discussed above and that it presented inedible qualities to the majority of the tasting panel and myself. In other words, it is chocolate dominated by off-flavors, such as vomit, found disgusting to the vast majority of people.[10]

This section explores and then explains this very wide range of quality for farm-to-bar and bean-to-bar chocolate from Kaua'i and the Big Island. The results of the tasting panel analysis, presented in table 8, corroborated most of my personal impressions of the best and worst bars. Based on the composite score category (derived from the four categories, each with a low of 1 to a high of 5), the average for the seventeen bars tasted was 2.63, while the standard deviation was 0.57. The high was 3.44 (between the categories of "Good" and "Very good" on the rubric), and the low was 1.36 (in the "Seriously flawed" category). In presenting the data, I reveal only the identities of the makers of bars that were considered above average by the panel, since I do not wish to harm the businesses of the makers of the worst bars, as many are still trying to improve their products. It should also be noted that not all makers from the two islands were represented in the tasting; thus, not

TABLE 8. Panel tasting results for seventeen bars from Kaua'i and the Big Island (taster $n = 9$)

SAMPLE #	MAKER	BAR NAME	BIG ISLAND							
			AROMA (1–5*) 15%	FLAVOR (1–5*) 45%	MOUTHFEEL (1–5*) 20%	OVERALL IMPRESSION (1–5*) 20%			COMPOSITE SCORE	
			Mean	Mean	Mean	Mean	SD	Rank	Mean	Rank
H1	Kuaiwi Farm	(dark chocolate, no label)	2.94	2.83	2.67	2.78†	0.67	7	2.81†	8
H2	Original Hawaiian Chocolate Factory	Dark Chocolate	3.22	3.11	2.78	3.06†	0.81	5	3.05†	4
H3	—	—	2.93	2.14	2.86	2.43§	0.79	10	2.46§	10
H4	—	—	2.11	1.89	2.56	2.22§	0.97	13	2.12§	13
H5	Sweet Dreams Hawaii	Lava Rock Chocolates (dark)	3.33	2.83	3.22	3.17†	1.06	3	3.05†	4
H6	Kahi Ola Mau Farm	70% Cacao Chocolate	3.11	3.50	3.11	3.11†	1.17	4	3.29†	2
H7	—	—	1.88	1.44	2.50	1.56ǁ	0.62	15	1.74ǁ	15
H8	—	—	2.13	2.31	2.75	2.31§	1.10	12	2.37§	12
H9	Red Water Café	Dark	2.50	3.00	2.75	3.06†	0.94	5	2.89†	6

KAUA'I

SAMPLE #	MAKER	BAR NAME	AROMA (1–5*) 15% Mean	FLAVOR (1–5*) 45% Mean	MOUTHFEEL (1–5*) 20% Mean	OVERALL IMPRESSION (1–5*) 20%			COMPOSITE SCORE	
						Mean	SD	Rank	Mean	Rank
K1	—	—	1.22	1.22	1.89	1.22‖	0.44	16	1.36‖	16
K2	Steelgrass Farm	70% Dark	2.44	3.00	2.44	2.56†	1.13	9	2.72†	9
K3	Steelgrass Farm	70% Dark with Sea Salt	3.83	3.39	3.28	3.44‡	1.24	1	3.44†	1
K4	Moloaʻa Bay	60% Dessert Chocolate	2.44	3.22	2.61	3.28†	1.03	2	2.99†	5
K5	Moloaʻa Bay	70% Smooth Dark	3.25	3.25	3.00	2.94‡	1.27	6	3.14‡	3
K6	Garden Island Chocolate	Dark Chocolate	3.17	2.72	3.13	2.67‡	0.71	8	2.86‡	7
K7	—	—	2.33	2.44	2.56	2.33§	0.50	11	2.43§	11
K8	—	—	2.11	1.78	2.50	1.72‖	1.25	14	1.96‖	14
		Average	2.64	2.59	2.74	2.58	0.92		2.63	
		Standard deviation (SD)	0.66	0.69	0.35	0.63	0.27		0.57	

Source: panel tasting.

* See table 6 for tasting rubric used.

†Good to Very good (≥ Average +1 S D): 3.21 to 3.44 / 3.20 to 3.44

‡Good (Average to +1 S D): 2.58 to 3.20 / 2.63 to 3.19

§Flawed (−1 S D to Average):1.95 to 2.57 / 2.05 to 2.62

‖Seriously flawed (≤ Average −1 S D): 1.22 to 1.94 / 1.36 to 2.04

being named does not mean a maker falls below average, as they could not have had bars within the tasting.

Some farm-to-bar and bean-to-bar producers on Kauaʻi and Hawaiʻi are doing an outstanding job making chocolate. My personal impressions as a craft chocolate aficionado were as follows. On Kauaʻi, Steelgrass Farm makes only exceptionally good single-estate chocolate. It is top notch within the world of craft chocolate. Another favorite of mine was Garden Island Chocolate. Their best chocolates—made from rare cacao beans from isolated groves on specific farms—came at the end of the chocolate tour and were truly exceptional, but they are not available as bars for the general public to consume.[11] On the Big Island, my favorite bean-to-bar tasting was at Kahi Ola Mau Farm in the northeast. They are making excellent chocolate even as they are getting started, although it is not available yet for general consumption as of the research visit in September 2017. Kuaiwi Farm south of Kona makes flavorful single-estate chocolate that is very good.

According to the composite index from the panel analysis data (table 8), Steelgrass Farm's 70% Dark with Sea Salt bar was the top-rated bar, at 3.44. Kahi Ola Mau Farm's 70% Cacao Chocolate bar was ranked second, at 3.29. Moloaʻa Bay's 70% Smooth Dark was the next highest rated, at 3.14. Sweet Dreams Hawaii Lava Rock Chocolates and the Original Hawaiian Chocolate Factory's Dark Chocolate tied for fourth, at 3.05. I considered all but one of these to be excellent chocolate according to the norms of craft chocolate.[12]

The three bars that I found to have the most offensive off-flavors were also the most poorly rated by the panel. Two of the three lowest-ranking bars were from two different farmers who had not yet fully commer-cialized their bars; they had composite index ratings of 1.96 and 1.74 and open-ended comments such as "puke aftertaste" and "very alkaline, bitter, alcohol, acetone." These farm-to-bar makers had been tinkering with their process for a few harvests and were giving away samples to friends and acquaintances rather than selling it. The one with the low-est score, at 1.36, is common in stores on Kaua'i. It received low panel scores—seven "seriously flawed" and two "flawed"—with remarkable qualitative comments, such as "Unacceptable. Like a raw, unfermented bean. Gives me a headache," "dog poop, sandy, sour, dirt," and "terrible finish that won't leave my mouth."[13] The question of how this and other subpar bars can be commercialized is one I consider below.

"Poisoning the Well": Bad Bars and the Proximate Reasons for Them

As the author of *American Terroir* has noted, "A lot of things have to go right to make good chocolate. A single misstep at any stage—poor fermentation, drying, roasting, or grinding—can ruin everything" (Jacobsen 2010, 253). Some makers I interviewed spoke to the strong off-flavor qualities in other local makers' chocolate. Jillian, a farm-to-bar maker, offered this vivid recounting:

> JILLIAN: I've had so many bad experiences with [eating locally made chocolate] that I'm like, "I don't even want to open my mouth for that, I just don't want to," which is why it really scares me because I'm sure that I'm not alone. The last time I ate that, I was aiming over the toilet [vomiting] and I don't want to do that again. . . . I was shocked, actually, because some of these guys are actually selling. Their [bar] was all nicely labeled and, "You can get it at this store and that store," and I was like, "God, that's too bad." . . . They probably started with a good product from the pod, we hope, but past that, they went downhill. I don't think they fermented properly, because that's where that putrid taste gets into the beans and there's just no way to get it out. No matter how much you roast it, it's just not going to go away. . . . You know like in Harry Potter, they had those jelly beans.

RG: Bertie Bott's.

JILLIAN: Yeah, earwax and stuff. These are like evil Hawai'i cacao jelly bean things. It's just like, "No, I don't want to do that, thanks." But [the bad experience] sticks. It really sticks.

A few were critical of a local maker who boasts of making and selling "the best chocolate in the world" at farmers' markets but who produces bad chocolate in their assessment.

Some insights from Dan O'Dougherty—trained in cacao growing and fermentation at the University of Hawai'i at Mānoa and now a leading world expert on specialty cacao fermentation who makes his livelihood traveling around to help farmers with fermentation—are useful here. In his talk on fermentation for quality cacao at the Chocolate Makers' UnConference in Seattle in November 2017, he briefly noted that for a long time he had tried to get some Hawaiian chocolate makers to stop "poisoning the well" of public perception of Hawaiian chocolate but eventually gave up. He half-jokingly conjectured that some makers must think that healthy food must taste awful, so that awful-tasting chocolate must be very healthy (O'Dougherty 2017).

As Jillian noted above, the source of the off-flavors is likely poor fermentation, since fermentation is challenging in small volumes. O'Dougherty (2017) argues that in his experience, it is impossible to produce consistency in cacao bean quality while fermenting with small volumes. This is perhaps the largest challenge of farm-to-bar production on the islands, since smaller single-estate makers often do not produce a high enough volume to yield consistent fermentation. To address this issue, researchers at the University of Hawai'i at Mānoa produced a short manual in 2009 on how small-scale growers or hobbyists can produce chocolate with as few as six pods; doing so requires a small-space controlled climate (like an insulated container), an inoculum of yeast, and a controlled heat source to keep the beans at 95°F for the first three days and at 113°F for days four through six (Bittenbender and Kling 2009).

I witnessed a much less controlled version of this process on many farms. Many farmers used ice chests and unpowered deep freezers and refrigerators as fermentation containers in an effort to contain heat and thereby raise the temperature. Some had an added heat source, but most were not monitoring the temperatures. In at least two operations where temperatures were being monitored, I saw that the beans were not reaching the temperatures needed to transform them, even according to the guide for small growers and hobbyists (Bittenbender and Kling 2009). All of this suggests that fermentation on many farms is not adequately transforming the beans to yield quality chocolate (see also chapter 3 on "landscapes of failure" for a landscape-level focus about why certain wine regions fail).[14]

Yet the chocolate-making process might also be responsible for some of the off-flavors. An important task of a chocolate maker is to sort the beans. The best craft chocolate makers meticulously sort through beans since there are always poor-quality beans due to the variability of the fermentation process (even the best batches meeting the highest grades of 95 percent well-fermented beans need to be sorted). Many makers of bad bars are likely not adequately sorting. Additionally, conching for long periods of time (twenty-four to seventy-six hours) has long been used as a way to mellow chocolate by volatilizing the acetic acid and other substances causing off-flavors (Jacobsen 2010). Although in my interviews I did not ask about conching time, it could be that conching for longer periods would allow for more off-flavor substances to volatilize.

Pinpointing the source of the off-flavor problem is impossible with the phenomenological approach I took to tasting; a more controlled approach is needed at the level of individual makers. For example, I had a bar from each of three different makers that all used the cacao beans of one of the biggest fermentaries on the Big Island. Of the three bars from these beans, one was excellent (with a symphony of blackberries, walnuts, and fudge), one mediocre, and one had a vomit flavor. The panel tasted only the first two of these bars, and the data also show that one ranked well above average (3.29) and the other, below average (2.22). While this observation hints that the chocolate-making process is the source of the problems, the bars were not necessarily sourced from the same exact batch of beans, so the fermentation could have been off for the worst bar. Thus, separating out the causes would require more research under more controlled conditions.

Another example of the difficulty of pinpointing the causes of low quality comes from tasting bars from the same maker over time. I found in a small store one maker's bars that were about six months old (not uncommon with chocolate, as its shelf life is generally one to two years) and found it to have some off-flavors. I later visited the maker again and tasted their most recently made chocolate, which was excellent and therefore a large improvement. During the interview I learned they had recently installed new chocolate-making equipment together with climate control (air-conditioning) in their factory. In addition, a dedicated, knowledgeable, and detail-oriented employee had recently taken over chocolate production. Since the change involved equipment, climate control, and the employee overseeing the making, it is impossible to know the contribution of each to the improved quality, but I suspect the detail-oriented employee makes the largest difference. Regardless of the precise contributions from each change in process, the combined effect suggests that more attention to detail can improve quality for individual makers.

*Connections with and Disconnections
from the Mainland Chocolate Movement*

In addition to the specific ways that production affects quality—as in the production of off-flavors through poor fermentation and/or by failing to remove what causes them as a result of poor manufacturing processes—my research unearthed a deeper reason for poor chocolate quality: lack of familiarity with mainland craft chocolate quality norms.

The main difference between the interviews with makers making excellent chocolate and those making bad chocolate was in the level of their connections to the mainland craft chocolate movement or to the two internationally known Hawaiian chocolate makers, Mānoa and Madre. With one exception, those makers with strong connections—acquaintance with well-known makers and/or regularly eating other craft chocolate—made excellent to very good chocolate. Interviewees also talked about the importance of these relationships. For example, staff from one of the craft chocolate leaders, Dandelion Chocolate in San Francisco, California, visited many makers in Hawai'i, and some makers noted that they have kept in touch and follow what Dandelion does.

Nathan, who is just starting out and makes excellent chocolate (one of the top rated by the panel and myself), shows the importance of these connections to the wider craft chocolate movement. He noted that "we buy chocolates, whenever we can, we taste them" in order to expand his palate. He is using roasting profiles from John Nanci—whom some call the grandfather of the craft chocolate movement—available online via the well-known blog/website *Chocolate Alchemy* (Nanci 2017). Nathan is also very detail oriented and keeps a log of his work, as a scientist does when conducting chemistry experiments. Thus, it is clear that building external references to quality, as well as training one's palate in this way, is an important foundation for excellent Hawaiian chocolate and works well when paired with a rigorous experimental approach.

Two other routes to producing excellent chocolate were suggested by individual cases. In one case a single-estate farm-to-bar operation contracts with one of the O'ahu-based makers to make their bars. This relationship allows the chocolate-making company—which has considerable experience with quality cacao beans from around the world—to provide feedback on the farm's beans and fermentation process. This sharing of expertise across the division of labor between maker and grower and across the value chain means that each can specialize in an area of expertise, and when this works synergistically, with considerable feedback between the two, the outcome is excellent.

Another route to excellence is deep experience with growing, fermenting, and making. One maker of excellent chocolate is composed of a wife-and-husband team deeply immersed in the place-based nature of cacao growing and fermentation in their region. The partners are constantly experimenting and know a great deal about different cacao varieties, where different varieties are grown and by whom, and how these specific varieties work with their fermentation and chocolate-making processes. In addition to growing their own cacao, they have deep relationships with many householders and have encouraged them to grow different varieties. They take fermentation very seriously, keeping a logbook of every fermentation they do, recording the source of the beans down to the field level, and experimenting with wine yeast inoculum, which they have used to good effect. One partner, Raymond, noted,

> At the start of every season we order about a dozen different types of wine yeast from different brew stores. And you know, the little packets? Red Star, Levine, different wine yeasts. And I experiment with the different yeasts and bacteria on the beans when they're fermenting. So we use banana leaves and then different strains of yeast. And I found over the years what strain of yeast works well on what variety [of cacao]. And then what type of fermentation box. So you've gotta control the temperature, humidity, and the oxygen. And it's all about the ferment. You mess up with the ferment, it's hard to recover from that chocolate, it could be full of off-flavors. And people may not even notice the off-flavors.

Thus, even though they are not deeply connected with the mainland craft chocolate movement, their deep experience in place and constant experimentation have yielded great results.

A couple of makers who made good but not outstanding chocolate were disconnected from the craft chocolate movement. Jesse, a chef who decided to make his own chocolate for his restaurant, makes good chocolate despite his lack of connection to the craft chocolate movement. His trained palate has allowed him to create good chocolate while being solidly within the foodie world of chefs. Similarly, when I asked Jillian, the farm-to-bar maker who noted the need to vomit upon tasting some local chocolate, what her favorite chocolate was, she named Lindt. She made moderately good chocolate, but a standard of Lindt means that she was missing out on many of the recent developments in the craft chocolate movement, especially for the most complex bars.

On the other hand, most of the makers of bad chocolate had weak or no connections to the craft chocolate movement, either on the mainland or to the two well-known Hawaiian makers. They do not actively try to expand or refine their palates through tasting other craft chocolate, and the results are off-flavors that would be completely unacceptable to mainland craft chocolate makers. But the explanation for how they can continue to sell vomit-flavored chocolate lies in the constant flow of new tourists. A consumer base with rapid turnover means that some chocolate makers do not need loyal consumers, which helps explain why there is little incentive for makers of bad bars to improve quality. For example, Ollie, one maker of what I would later taste and characterize as a low-quality bar, noted, "That's like 99 percent of sales, is brand-new customers," even though he did not want it to be that way. He also noted that there is always enough demand for his bars and everyone else's: "Everyone sells out. Anyone who's making chocolate in Hawai'i . . . it's all sold."

Conclusion

Exciting developments are happening in Hawaiian cacao and chocolate. These include expanded plantings of cacao, chocolate agritourism, and a growing farm-to-bar and bean-to-bar movement. My concern for the industry is that very uneven quality will restrict its success. For this reason, the politics of quality is an important point to end on. Leissle (2018, 159) has noted that "one way to understand the politics of quality is to consider that its governance—how quality is measured, valued, and enforced, and by whom—is not a bottom up endeavor." While true for the vast majority of places where cacao grows—where quality standards created somewhere beyond the farm dictate practices to some extent, as in many other Global South-North commodity chains (Galt 2014)—the governance of cacao and chocolate quality in Hawai'i *is* largely a bottom-up endeavor. Farm-to-bar makers control the whole value chain, and small bean-to-bar makers do not have the economic leverage to dictate quality standards for the local growers of the cacao beans they purchase. As such, the Hawaiian context is very unique and offers interesting possibilities in the governance of quality.

At the level of individual makers, experimentation—which abounds in the Hawaiian chocolate world—is essential, *but*, in order for many makers to come near the norms of craft chocolate, innovation must be paired with a trained and constantly expanding palate informed by what is occurring in the mainland and international craft chocolate movements. All locally made Hawaiian chocolate retails at about the same price. Currently, for the bad and mediocre bars, the only source of value that can justify the expense (in the mind of discerning consumers) is novelty. This novelty will not be converted into customer loyalty if the novel experience is negative, and a negative experience by a consumer based on one bar of local chocolate can taint the image of the whole industry for that consumer and whomever they tell.

As a value proposition, then, Hawaiian chocolate needs to provide consistent quality—across all makers—to continue to attract all people who try it and to build a returning consumer base. Many individual makers provide consistently high quality, but with one-fifth of bars from small farm-to-bar and bean-to-bar makers having off-flavors, the consumer has a highly variable, hit-or-miss experience with Hawaiian chocolate: some will taste only great chocolate and others will taste only bad chocolate, and those more keen on tasting chocolate will likely taste both. This means a large proportion of chocolate consumers will come away from their chocolate-eating experience with only a negative impression—or at best a decidedly mixed one. This variation will make it impossible for Hawaiian chocolate to create a unified image of good quality.

Changes at the collective level are essential if creating a unified image of good quality is important to makers in Hawai'i. Chocolate makers can consider using their professional organizations to enhance quality for all makers. Indeed many chocolate makers on the same island are isolated from one another, while a cluster mentality focused on a collaborative economy—recognizing shared interests and acting strategically to advance them—could serve everyone well, as it has in wine regions (see chapter 2, on successes, and chapter 3, on failures). For example, in analyzing Napa Valley's success, Anil Hira and Tim Swartz (2014, 51) conclude, "Terroir is just a starting point for understanding how quality is created in wine; the rest depends upon human agency, particularly the efforts of entrepreneurs, to work together to develop the technological breakthroughs that give them a comparative advantage. Behind these breakthroughs and the development of a regional brand are social capital and institutions." At the least, a starting place could be creating a collectively agreed upon standard for off-flavors and then finding a way to implement it. While associations cannot force makers to enhance quality, there might be ways to work together to identify the sources of the off-flavors (and poor temper) and thereby rectify them. Indeed a new method of microbatch fermentation has been published and might provide a solution to the off-flavors produced in small-scale fermentation (Bittenbender et al. 2017). Additionally, many chocolate makers on the islands need a broader understanding of the other quality norms of the mainland and international craft chocolate movement, since this is the context in which discerning customers are increasingly immersed.

Lastly, greater synergies could be created between chocolate makers in Hawaiʻi and in other locations. Some mainland makers bring loyal customers along on their sourcing trips and other excursions to producing regions. This gives their eager customers an opportunity to see cacao growing and to understand some of its production processes and challenges, likely further enhancing their engagement in craft chocolate. Mainland bean-to-bar makers could highlight tours in Hawaiʻi as one of the few cacao trips that does not require a passport. Yet, in doing so, they would want to be sure that their consumers would experience high-quality chocolate, which is another reason to address collective quality control. Lastly, the nonprofit Multinational Exchange for Sustainable Agriculture (MESA) has brought Ecuadorian coffee farmers to Hawaiʻi for knowledge exchange, and a similar type of exchange could be established between Hawaiian cacao farmers and Latin American cacao farmers as a way of learning from one another. Many Latin American countries have long histories of growing and fermenting cacao and making chocolate. Overall, the potential for farm-to-bar and bean-to-bar craft chocolate to link fermented landscapes of cacao to localized livelihoods in Hawai'i is quite large; my hope is that through the strategies discussed above, more benefits can be realized.

Notes

I am deeply grateful to all of the craft chocolate makers who participated in this research, especially Maddy Smith, who introduced me to many chocolate makers on the Big Island. Many thanks also go to my family, Eve and Paige, who made the fieldwork even more fun and memorable than usual. Funding for the project came from the MacArthur Foundation Endowed Chair in Global Conservation and Sustainable Development position that I hold at the University of California, Davis.

1. I use the term "Big Island" to refer to Hawai'i Island throughout this chapter, and I use "Hawai'i" alone when referring to the entire state.

2. In recent years "bean-to-bar chocolate maker" has become a common term for referring to chocolate makers who buy dried cacao beans as their raw material for making chocolate (Williams and Eber 2012; Giller 2017). This differs from chocolatiers and confectioners who buy premade chocolate (commonly called couverture, in block or disc form) and melt it to create their chocolate products. Here I use "farm-to-bar chocolate maker" to refer to a subset of bean-to-bar makers who also grow their own cacao.

3. This pattern mostly remains in place today, but some changes are occurring as chocolate making is increasing in countries of origin (Leissle 2018).

4. There are ongoing efforts to create formal standards and have them adopted. Indeed efforts are under way by the Fine Chocolate and Cacao Institute, led by Carla Martin (pers. comm., January 20, 2018), to spread a standard way of evaluating dried specialty cacao beans within a variety of cacao-producing and chocolate-making countries. Additionally, some makers, such as TCHO, have set up their own standards by using collaborative software like Cropster (used initially in quality coffee) to calibrate palates of makers and growers across large distances.

5. An example of this mixed take is TCHO's original chocolate flavor wheel. TCHO is one of the larger craft chocolate makers with national distribution and is well respected within the industry, even if looked upon by some as being too large to be craft. TCHO's original flavor wheel describing its six dark chocolates included the flavors "floral" and "earthy." While bars with the other flavors—"chocolatey," "nutty," "bright" (citrusy), and "fruity"—have been relatively easy to find from 2014 and on, I had never seen either "floral" or "earthy." I asked Laura Sweitzer, TCHO Source program manager, about this, and she said that those two had proved to be the least

popular so the company no longer produced them (pers. comm., March 28, 2016). Additionally, these flavor preferences vary culturally, such that in Japan, with its own growing bean-to-bar movement, earthy and umami flavors are more normal, as evidenced when comparing Dandelion Chocolate's bars made in San Francisco for the U.S. consumer and those made by Dandelion in Tokyo for the Japanese consumer.

6. There are many other norms in the craft chocolate industry that are not related to flavor and chocolate products themselves. These include an openness to sharing information, as well as attempts at transparency in sourcing, since industrial chocolate makers are notoriously opaque about sourcing; the prices paid to farmers; and the labor conditions on farms, in the wake of exposés about child slavery in Côte d'Ivoire and Ghana, which produce half of the world's cacao beans (Off 2008; Bertrand 2011).

7. The tours were so similar that one farm felt compelled to note in its brochure that it is the original. Both were well attended.

8. All names are pseudonyms to preserve confidentiality of the interviewees.

9. Bernardini (2015) rigorously tasted chocolate from all of the world's makers that he could find, and he detailed all of their flavor profiles. He came up with his lists of the world's best, including a list of dark chocolate bars. Having personally eaten six of the fifteen bars he considered the best, I tend to agree that they are all top notch.

10. As noted in chapter 1, disgust is a common, and likely protective, human reaction to the process of decay; that some of the less-than-savory flavors of the decay process make it into, or even dominate, some chocolate is perhaps not surprising, even though it is uncommon among professional craft chocolate makers on the mainland. Other than vomit-like off-flavors for about a fifth of the bars, some makers' bars were subpar due to temper. The lack of a snap suggests that tempering processes have not been perfected by a number of makers, likely due to the challenge of a lack of cool temperatures (described earlier) and, potentially, lack of tempering experience.

11. Garden Island Chocolate's mainstream bars, found in grocery stores, are mediocre from a flavor perspective. The maker adds Ecuadorian cacao beans to the fine-quality Kaua'i-grown beans they harvest, which, to me, overpowers the local chocolate flavor, making the bars unremarkable. The tasting panel confirmed this, with Garden Island's store-bought bar falling squarely within the middle of the quality range.

12. The divergence is with Sweet Dreams Hawaii Lava Rock Chocolates, which to me has a mild off-flavor of vomit; this was not detected by the panel. Interestingly, one interviewee talked about this exact problem for this maker, noting that "[the maker] says he has a hard time getting a consistent product and a lot of his customers say it tastes like vomit, he was saying. It wildly varies."

13. I found two other bars to have predominant off-flavors, but these were not found by a majority of the panel tasters, so I do not mention them here.

14. A friend in the chocolate world told me the story of a chocolate-making friend who ordered a batch of cacao beans from one of the larger fermentaries on the Big Island. The beans looked poorly fermented, to the extent that there was a fear that molds had likely developed in them; they did indeed produce low-quality chocolate.

References

Afoakwa, Emmanuel Ohene, Alistair Paterson, Mark Fowler, and Angela Ryan. 2008. "Flavor Formation and Character in Cocoa and Chocolate: A Critical Review." *Critical Reviews in Food Science and Nutrition* 48 (9): 840–57. https://doi.org/10.1080/10408390701719272.

Bernardini, Georg. 2015. *Chocolate—The Reference Standard: The Chocolate Tester 2015.* Translated by Brigitte Foley. Bonn: Forster Media.

Bertrand, William E. 2011. *Oversight of Public and Private Initiatives to Eliminate the Worst Forms of Child Labor in the Cocoa Sector in Côte d'Ivoire and Ghana.* Project spearheaded by the Payson Center for International Development and Technology Transfer, Tulane University, New Orleans.

Bittenbender, Harry C. "Skip," and Erik Kling. 2009. *Making Chocolate from Scratch.* Food Safety and Technology bulletin FST-33. Honolulu: University of Hawai'i.

Bittenbender, Harry C., Loren D. Gautz, Ed Seguine, and Jason L. Myers. 2017. "Microfermentation of Cacao: The CTAHR Bag System." *HortTechnology* 27 (5): 690–94.

Cheng, Martha, and Katrina Valcourt. 2015. "Hawai'i Chocolate: The Everything Guide to Local Chocolate." *Honolulu Magazine*, December 7, 2015.

Coe, Sophie D., and Michael D. Coe. 2013. *The True History of Chocolate.* 3rd ed. New York: Thames and Hudson.

Crawford, David Livingston. 1937. *Hawaii's Crop Parade.* Honolulu: Advertiser Publishing Company.

Ernst, Cheryl S. 2017. "University of Hawai'i Cacao among World's Best." University of Hawai'i at Mānoa, November 8, 2017. http://manoa.hawaii.edu/news/article.php?aId=8904.

Fleming, Kent, Virginia Easton Smith, and H. C. "Skip" Bittenbender. 2009. *The Economics of Cacao Production in Kona.* Oahu: University of Hawai'i, Mānoa.

Galt, Ryan E. 2014. *Food Systems in an Unequal World: Pesticides, Vegetables, and Agrarian Capitalism in Costa Rica.* Tucson: University of Arizona Press.

Galt, Ryan E. Madeline Weeks, Nicholas I. Robinson, and Angela M. Chapman. In preparation. "Chocolate's Transnational Terroir: Connecting Production to Consumption in the Specialty Cacao-Craft Chocolate Value Chain." For submission to *Geographical Journal.*

Giller, Megan. 2017. *Bean-to-Bar Chocolate: America's Craft Chocolate Revolution*. North Adams MA: Storey.

Gomes, Andrew. 2010. "Hoping for a Sweet Harvest." *Honolulu Advertiser*, January 25, 2010.

Hira, Anil, and Tim Swartz. 2014. "What Makes Napa Napa? The Roots of Success in the Wine Industry." *Wine Economics and Policy* 3 (1): 37–53. https://doi.org/10.1016/j.wep.2014.02.001.

Hoehn-Weiss, Manuela. 2018. "FCIA Business Survey of the Fine Chocolate Industry." Presentation at Fine Chocolate Industry Association's Elevate Chocolate meeting, San Francisco, Winter 2018.

ICCO (International Cocoa Organization). 2019. "Fine or Flavour Cocoa." Accessed May 7, 2019. http://www.icco.org/about-cocoa/fine-or-flavour-cocoa.html.

Jacobsen, Rowan. 2010. *American Terroir: Savoring the Flavors of Our Woods, Waters, and Fields*. New York: Bloomsbury.

Klassen, Seneca. 2016. "Cacao and Forest Conservation: Is It possible?" Presentation at the Northwest Chocolate Festival, Seattle WA, November 13, 2016.

Leissle, Kristy. 2013a. "Invisible West Africa: The Politics of Single Origin Chocolate." *Gastronomica: The Journal of Food and Culture* 13 (3): 22–31. https://doi.org/10.1525/gfc.2013.13.3.22.

———. 2013b. "What's Fairer Than Fair Trade? Try Direct Trade with Cocoa Farmers." *YES! Magazine*, October 4, 2013.

———. 2018. *Cocoa*. Medford MA: Polity Press.

Martin, Carla, and Kathryn E. Sampeck. 2015. "The Bitter and Sweet of Chocolate in Europe." Paper presented at the Social Meaning of Food Workshop, Budapest, Hungary, June 16–17, 2015.

Motamayor, Juan C., Philippe Lachenaud, Jay Wallace da Silva e Mota, Rey Loor, David N. Kuhn, J. Steven Brown, and Raymond J. Schnell. 2008. "Geographic and Genetic Population Differentiation of the Amazonian Chocolate Tree (*Theobroma cacao* L)." *PLoS One* 3 (10): e3311.

Nanci, John. 2017. "Ask the Alchemist #200." *Chocolate Alchemy*, April 6, 2017. http://chocolatealchemy.com/blog/2017/4/5/ask-the-alchemist-200?rq=roasting%20profile.

Nanci, John. n.d. "Conching and Refining." *Chocolate Alchemy*. Accessed May 28, 2018. http://chocolatealchemy.com/conching-and-refining/.

Nesto, Bill. 2010. "Discovering Terroir in the World of Chocolate." *Gastronomica* 10 (1): 131–35.

Nieburg, Oliver. 2016. "Premium Chocolate 'Leg Up': How to Win Fine Flavor Cocoa Status." *Confectionery News*. Last updated May 11, 2016. https://www

.confectionerynews.com/Article/2016/05/10/Everything-you-need-to-know
-about-fine-flavor-cocoa.

Notman, Nina. 2015. "Well-Tempered Chocolate." *Chemistry World*, December 1, 2015.
https://www.chemistryworld.com/features/well-tempered-chocolate/9200.article.

O'Dougherty, Daniel. 2017. "Cacao Fermentation." Presentation at the Northwest
Chocolate Makers UnConference, Seattle WA, November 9, 2017.

Off, Carol. 2008. *Bitter Chocolate: The Dark Side of the World's Most Seductive Sweet*.
New York: New Press.

Rice, Robert A., and Russell Greenberg. 2000. "Cacao Cultivation and the Conservation of Biological Diversity." *Ambio* 29 (3): 167–73.

Saltini, Rolando, Renzo Akkerman, and Stina Frosch. 2013. "Optimizing Chocolate
Production through Traceability: A Review of the Influence of Farming Practices
on Cocoa Bean Quality." *Food Control* 29 (1): 167–87.

Schnell, Raymond J., Cecile T. Olano, J. S. Brown, Alan W. Meerow, Cuauhtemoc
Cervantes-Martinez, Chifumi Nagai, and Juan C. Motamayor. 2005. "Retrospective
Determination of the Parental Population of Superior Cacao (*Theobroma cacao* L.)
Seedlings and Association of Microsatellite Alleles with Productivity." *Journal of
the American Society for Horticultural Science* 130 (2): 181–90.

Seguine, Edward, and Lyndel Meinhardt. 2014. "Cacao Flavor through Genetics—
Anatomy of Fine Flavor." *Manufacturing Confectioner* (November): 25–30.

Sethi, Simran. 2015. *Bread, Wine, Chocolate: The Slow Loss of Foods We Love*. San
Francisco: HarperOne.

Smillie, Eric. 2016. "Fermentation May Be More Vital for the Flavor of Chocolate
Than It Is for Wine." *Newsweek*, March 24, 2016.

Stanislawski, Dan. 1970. *Landscapes of Bacchus: The Vine in Portugal*. Austin: University of Texas Press.

Sukha, Darin A., David R. Butler, E. A. Comissiong, and Pathmanathan Umaharan.
2014. "The Impact of Processing Location and Growing Environment on Flavor
in Cocoa (*Theobroma cacao* L.)—Implications for 'Terroir' and Certification-
Processing Location Study." *Acta Horticulturae* 1047:255–62. https://doi.org/10
.17660/ActaHortic.2014.1047.31.

Trubek, Amy B. 2008. *The Taste of Place: A Cultural Journey into Terroir*. Berkeley:
University of California Press.

Williams, Pam, and Jim Eber. 2012. *Raising the Bar: The Future of Fine Chocolate*.
Vancouver BC: Chelsea Green.

Kombucha Culture

9

An Ethnography of Fermentos in San Marcos, Texas

Elizabeth Yarbrough, Colleen C. Myles, and Colton Coiner

Kombucha, an effervescent fermented tea beverage, has experienced tremendous growth in popularity in the United States since the mid-2000s. Sandor Katz (2012, 167), the self-titled "fermentation revivalist" and renowned expert on all things fermented, proclaims that "no other ferment even approaches kombucha in terms of its sudden dramatic popularity." According to a 2015 research study, the international market for kombucha is projected to increase from $600 million in 2015 to $1.8 billion in 2020 (PR Newswire 2016). Bottled and sold as a health drink, kombucha is promoted for its refreshing qualities and nourishing health benefits. Although this beverage has roots that can be traced back thousands of years, as recently as the mid-1990s most Americans were not familiar with the drink. Hardly any bottled kombucha was sold in the United States prior to the boom; it was available only in select specialty health food stores and from those few people who were brewing their own tea at home.

Today in Central Texas kombucha seems to be bubbling up everywhere: major supermarket chains, gas stations and corner stores, farmers' markets, coffee shops, restaurants, bars, and increasingly in residential home kitchens. There are now numerous commercial brands to choose from, such as GT's, Health-Ade, and Live kombucha soda, which have a nationwide market, in addition to multiple emerging local brands such as Buddha's Brew, Kosmic Kombucha, and Wonder-Pilz being sold in and around Austin and San Marcos, Texas, at specialty shops and farmers' markets. Why has this fizzy tea beverage become so trendy? And, more important to this study, why are so many people deciding to take on the practice of brewing their own kombucha at home? As a resident of Central Texas, I noticed this sweeping trend of homebrewing kombucha because, as I looked around me, I found many of my friends and acquaintances in San Marcos were brewing this fermented beverage at home.[1] Indeed I even took up the practice myself. There have been many reports published chronicling the increase in corporate kombucha sales (PR Newswire 2016; Carr 2014) and countless testimonies touting kombucha's health benefits (Frank 1995; Crum and LaGory 2016), but there is little to nothing chronicling what motivates people to brew this beverage at home in their own communities. Thus, the research project behind this chapter has employed an ethnographic methodological approach to explore what is driving this developing phenomenon at the local scale. It also leveraged my own connections to the homebrew culture in San Marcos. Through interviews, focus groups, and purposeful participation in and observation of kombucha brewing practice, this research has aimed to uncover the shared values, motivations, and inspirations of kombucha culture and demonstrate the community-building possibilities of this practice.

The Cloudy History of Kombucha

Kombucha is both a global and a historical beverage that has been brewed by many different communities throughout the world and across different eras of history. The precise origin of kombucha is contested, although the general agreement is that kombucha has its roots in eastern Asia. According to Hannah Crum and Alex LaGory, authors of the popular website Kombucha Kamp and of arguably the most comprehensive, contemporary resource on kombucha brewing, *The Big Book of Kombucha*, kombucha tea originated in the Manchurian region of northeastern China around 220 BCE. This "tea of immortality" was rumored to be reserved exclusively for the consumption of Emperor

Qin Shi Huang. The tea was not commonly consumed by the general population until much later; written records reference kombucha consumption for the masses during the Tang dynasty (618 to 907 CE). Still other stories claim kombucha began in Japan. One legend has a Dr. Kombu treating the Japanese emperor with the beverage around 415 CE (Frank 1995). As is the case with many folk recipes and remedies, the exact history or precise geography of kombucha can be difficult to pin down. More important than debating exactly where and when kombucha came from is appreciating the traditional lore and legends surrounding this beverage and recognizing the long relationship humans have had with kombucha across space and time. Just as physical materials were exchanged along ancient trade routes, so too were ideological beliefs and cultural traditions (Pryor 1996). Kombucha traversed the Silk Road, spreading across networks throughout Mongolia, Russia, and India and even reaching as far as the Middle East, likely in the flasks and canteens of traveling merchants (Hobbs 1995).

Eventually kombucha became popular and quite common in Russia, spreading first to Germany and into the rest of eastern Europe by 1852. The first documented scientific experiments concerning kombucha were carried out by curious Germans and Russians in the latter half of the nineteenth century. Dr. Nikolay Kirilov, a Russian ethnographer and student of Tibetan medicine, was the first to publish on the health benefits of drinking kombucha (Kaufmann 2013).

Up until World War II, and for the following couple of decades as well, hundreds of scientific research papers were published, predominantly by Russian and German doctors and scientists, documenting kombucha as a treatment for a variety of ailments (Crum and LaGory 2016). During the 1950s kombucha had a dramatic, albeit short-lived, boost in popularity with Italians. Soldiers stationed in Russia during World War II returned home with the "Chinese fungus," as kombucha was known, and an odd but trendy custom evolved. According to popular lore, scobys—from the acronym SCOBY, for symbiotic culture of bacteria and yeasts, the essential kombucha "mother"—were only to be passed on to friends on Tuesdays, and if the friend was successful in brewing, three wishes would be granted to the original brewer. This notion was memorialized in a pop song by Renato Carosone (Pryor and Holst 1996). The brief Italian boom of kombucha is just one of many colorful examples of people's enthusiastic experiences with kombucha in the past.

Out of the turmoil and sweeping change of the 1960s, a new kombucha following appeared on the San Francisco hippie scene. This was probably the first time kombucha was popular outside of a few immigrant hands in the United States (Pryor and Holst 1996). The kombucha clique stayed relatively quiet over the following decades; a few articles and books concerning kombucha's health benefits appeared in print, but none gained much traction in the popular press and kombucha largely remained unknown. It has been rumored that Ronald Reagan was a religious kombucha consumer during the late 1980s, drinking as much as a liter daily to combat his cancer diagnosis (Crum and LaGory 2016). By the 1990s and through the 2000s the practice of kombucha homebrewing in the United States was growing, spearheaded by notable author Günther Frank and activist Betsy Pryor. Today's fermentation guru Sandor Katz has also noted the dramatic rise in popularity of the beverage in his path-breaking book, *The Art of Fermentation*.

As consumers begin to be more concerned with health, commercial kombucha producers have responded. Today kombucha is mass produced for an expanding global market, bottled and sold as a health drink and functional food, that is, a food product, or beverage, that provides wellness beyond basic nutrition, via added vitamins, enzymes, and probiotics (Scott and Sullivan 2008). The largest share of this market belongs to GT's Kombucha, founded in 2001 by GT Dave, inspired by his mother's astonishing recovery from cancer, a recovery she credited to her daily consumption of kombucha. Within ten years two other major brands—Wonder Drink and High Country—joined the national market, and kombucha was undeniably on the commercial beverage map. As noted, an article in PR Newswire (2016) has predicted astronomical growth in the global commercial market, with profits soaring to $1.8 billion by 2020.

While there is an increasing amount of research being conducted on the biochemical composition and nutrition components of kombucha, there are also many personal testimonies touting kombucha as a cure for cancer or paralysis. However, no research to date suggests kombucha as a medical treatment for any specific ailments; rather, the evidence supports the notion of kombucha as a healthy beverage option that can be an integral part of a healthy lifestyle that in turn may lower overall vulnerability to poor health and disease.

How to Brew

Kombucha can be made a few different ways, but the basic recipe is described here as it was described by the study participants. Most people involved in this research project brew and store their kombucha in large glass containers, typically one- or two-gallon jars, and then transfer the finished product to sixteen-ounce swing-top bottles. The process begins with freshly brewed black tea. It is then sweetened with sugar, and a scoby is added with a bit of "starter tea" (finished kombucha) or vinegar, to keep the pH balanced. The jar is then lightly sealed with a cloth secured by a rubber band and left to ferment in a place where the process will not be disturbed by sunlight or high levels of activity. These handwritten directions (fig. 36) were originally scribed by a homebrewer named Ana and given to homebrewer Alex, who photocopied them and passed them on to me.

What Is a SCOBY?

SCOBY is the acronym for symbiotic community (or culture, or colony) of bacteria and yeast. This term, usually presented in lowercase letters, was coined by Len Porzio, author of one of the earliest listservs dedicated to kombucha, Balance Your Brew, to distinguish kombucha tea from the kombucha culture (Crum and LaGory 2016). In biology a symbiotic relationship exists between two different life forms living

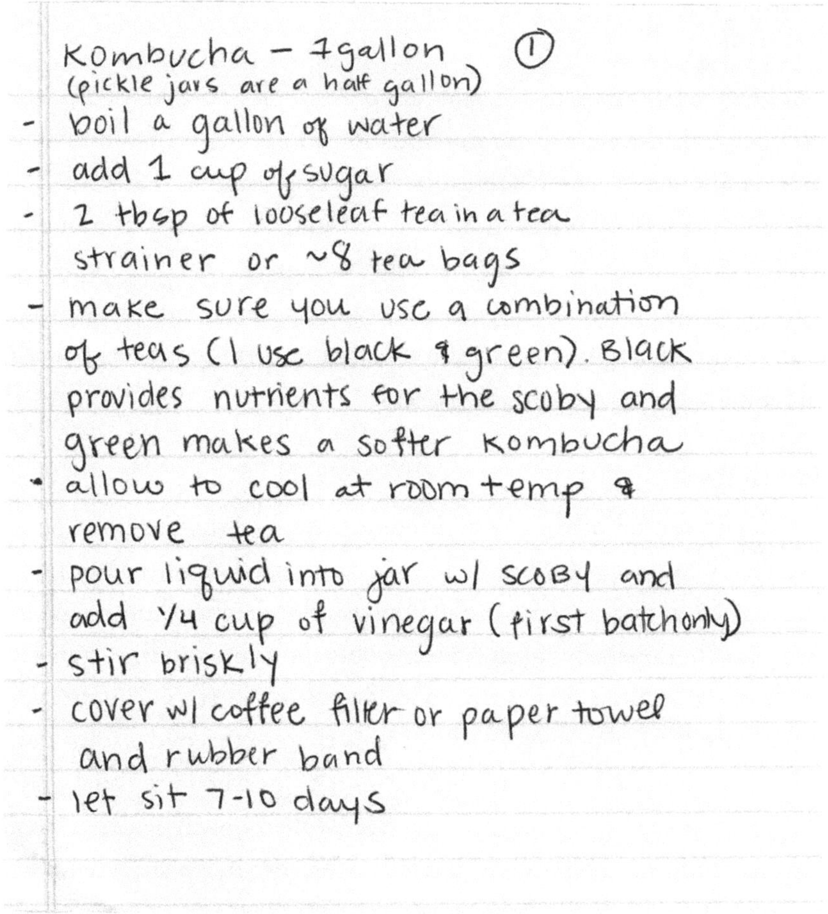

Kombucha — 7 gallon ①
(pickle jars are a half gallon)
- boil a gallon of water
- add 1 cup of sugar
- 2 tbsp of looseleaf tea in a tea
  strainer or ~8 tea bags
- make sure you use a combination
  of teas (I use black & green). Black
  provides nutrients for the scoby and
  green makes a softer kombucha
- allow to cool at room temp &
  remove tea
- pour liquid into jar w/ scoby and
  add 1/4 cup of vinegar (first batch only)
- stir briskly
- cover w/ coffee filter or paper towel
  and rubber band
- let sit 7-10 days

FIG. 36. Handwritten kombucha instructions, passed on from one friend to the next, similar to how a scoby is passed along. Recipe by Ana Arguello. Used with permission.

together, which, as in the case of the scoby, is mutually supportive. A scoby is scientifically defined as a cellulosic pellicle with embedded bacteria and yeast living in symbiosis (Marsh et al. 2014). There is some debate among fermenters as to what the *C* in scoby stands for, variations of which can include colony, culture, or community. *Colony* is the least accepted word because, by definition, a colony is a group of the same kind of organisms living together, which is not the scenario with a scoby. *Culture*, on the other hand, refers to the substrate in which the bacterial

IMPORTANT   ②
- make sure everything is very clean or you will kill the SCOBY
- do not store in cabinet- likes air flow
- SCOBY can sink or float - both okay
- white bubbles on SCOBY are normal
- dont mix anything other than sugar + tea w/scoby ↑ +water

2F - flavoring
- move kombucha from scoby jar to different jar
- add whatever flavoring ingredients you want to try (fresh or frozen fruits, herbs, extracts, whatever)
- let sit for another few days w/cap or no cap, up to you, cap makes it fizzy

species are living and does not recognize the organisms present. Many in the brewing community thus prefer the term *community* because the connotations associated with this word more accurately describe the nature of the scoby. Merriam-Webster's dictionary defines *community* as an interacting population of various kinds of individuals (species) in a common location, which exactly describes the relationship between the various yeasts and bacteria species within the floating, gelatinous mat. An additional definition of the word *community* is a feeling of fellowship with others, as a result of sharing common attitudes, interests, and goals. This definition resonates with one of the key themes that this project is exploring—the sense of community gained through kombucha brewing.

There are also several misnomers associated with the scoby. For example, scobys are sometimes called mushrooms, but since a mushroom is the fruiting body of a fungus (Stamens 2005), this is clearly an incorrect characterization. Another common but imprecise name for the scoby is simply "the mother." As a successful round of fermentation occurs, the scoby creates a new layer on top of the original mother, creating the next generation. As a scoby is constantly reproducing itself, it is more like a clone than one half of a parent partnership. That said, I, as well as others, do find myself calling the freshest layers of the scoby that develop on top of the original disk "babies," which is also technically inaccurate. A discussion of the semantics of scobys, cultures, brewing practices, and components is important because the varying names used to refer to the scoby can elicit different feelings and implications in brewers and thus can influence the relationship between the brewer and the brew (see chapter 1 and chapter 12 for discussions of the coevolution of humans and microbes).

In addition to the composition of the scoby and the brewer's understanding of it, the source of the material is also important. In other words, it is useful to consider how brewers procure their scobys. Several routes exist, including ordering a dehydrated scoby online to be shipped through the mail or by patiently growing one from a bottle of unflavored, store-bought kombucha. However, these techniques lack personal connections, do not forge any new relationships, and cost money. Instead scobys are often passed from brewer to brewer, an exchange made possible by the fermentation process itself. As the homebrew fermentation process undergoes multiple iterations, the scoby becomes a multilayered structure and can be gently peeled apart without any damage occurring to the organisms. This new scoby can then be given to a friend, thus freeing up space in your original jar and enabling someone else to make kombucha at home. In turn, the new scoby will grow larger, and the cycle of multiplication, division, and sharing endures.

The traditional technique of passing scobys from person to person is at the heart of kombucha culture and, as the preferred means of scoby acquisition by brewers, is the focus of this project. Heather Paxson (2008, 25) describes the "social character of microbes: [as] natural flora and fauna, they materialize as specific communities within ecologies of human practice. To speak doubly of cheese cultures—bacterial and human—is no idle pun." Similarly, this chapter describes how the practice of passing cultures becomes a culture in and of itself.

As time passes and the scoby grows, the microorganisms feed off the caffeine and sugars from the tea, converting them to acids and alcohols with a bubbly carbon dioxide by-product. A shorter fermentation time will yield a milder, sweeter, flat tea, and as brew time is extended, the drink becomes fizzier and more sour tasting or vinegary. The amount of time it takes to ferment depends on several factors, with air temperature and personal taste preferences being the main determinants, but the tea is usually complete within ten to twenty days of brewing on a countertop or bookshelf in the ambient atmosphere of a Central Texas home. At this point, brewers may choose to bottle the kombucha and refrigerate the tea (cold temperatures nearly halt the fermentation process), or they may elect to do a second round of fermentation, which increases carbonation and is typically the step when flavors are added. Other approaches to making kombucha exist, such as the continuous brew method, but most of the participants involved in this study followed the steps outlined above.

Site and Situation: San Marcos, Texas

This site of this project was San Marcos, Texas, the seat of Hays County, situated about thirty miles south of Austin along the I-35 corridor and the Balcones Escarpment, a geologic fault zone where the Gulf Coastal Plains ecological region transitions into the Texas Hill Country. The rich, fertile soil of the Blackland Prairie to the east traditionally supported a cotton-based economy whereas cattle ranching reigned supreme across the rocky limestone hills of the Edwards Plateau. The karst topography of the Edwards Plateau has defined the aquifer system and created the artesian wells and spring-fed rivers characteristic of the Central Texas region. According to the underwater archaeologist Joel Shiner (1983), the immediate area of San Marcos Springs, bubbling up beneath Spring Lake, has a history of human habitation that goes back to the Clovis people.

For three consecutive years (2013, 2014, and 2015) the U.S. Census Bureau named San Marcos the fastest-growing city in the country. Estimates put the 2016 population at around 60,684, which is more than double the figure from 1990 (U.S. Census Bureau 2015). This remarkable increase is representative of a regional pattern of explosive population growth. The Austin–Round Rock–San Marcos metro area is booming for several reasons, including a robust economic culture that remained relatively unscathed during the financial crisis that began in December 2007, as well as a business-friendly regulatory context that entices new companies to invest in the area (e.g., the Amazon Fulfillment Center established in San Marcos in 2016). Retail tourism plays a primary economic role in the city, attracting shoppers to the downtown square and to the largest outlet shopping center in the United States: the combined Tanger Outlets and San Marcos Premium Outlets. In addition, San Marcos is home to Texas State University, the fourth-largest public university in Texas (more than thirty-eight thousand students) and one that is expanding its enrollment and footprint every semester (Office of Institutional Research 2017).

San Marcos experiences mild winters and long, hot summers, typical of the humid subtropical climate designation (Dixon and Bray 2010). The outdoor enthusiast has many options when it comes to local parks and natural areas, with miles of beautiful trails to hike or bike across more than twelve hundred acres of protected spaces (San Marcos Greenbelt Alliance 2015). The beautiful San Marcos River flows through town with plenty of public access for locals, students, and visitors to enjoy swimming, snorkeling, paddling, and fishing. Many residents are committed to protecting the unique and sensitive river, which is home to several endangered species, including *Zizania texana* (Texas wild rice), which grows only in the upper portion of the San Marcos River. Love of the river and safeguarding the beautiful land around town are common traits of many homebrewers.

The culture of the San Martians (as residents of San Marcos call themselves) is vibrant, colorful, and passionate. The arts scene is alive and thriving; many public murals are being commissioned around town on formerly blank walls, new art galleries are opening and showcasing a variety of media, and pop-up events are becoming the norm. The local music scene is also flourishing, along with other forms of entertainment such as stand-up comedy and slam poetry readings. There is a "hippie" vibe around town, likely due to the youthful influence of college students and idealists who call this town home. Young people are some of the earliest adopters of new ideas and technologies, especially those marketed as "green." Many residents, including those who participated in this study, are among the "alternative" or progressive camp: those who rally for environmental causes as well as social justice ones. Many residents value healthy lifestyles, as is evident from the health food, smoothie, and supplement/vitamin stores; the Healthy Living section at the large regional chain grocery store; the expanding number of yoga studios, gyms, and workout groups; and a general enthusiasm for living a more authentic life.

Theoretical Framework

While there is limited literature and little prior research on the social aspects of kombucha, there are copious technical publications on the topic, generally in fields such as microbiology and nutrition science. These technical, microbiological studies of kombucha are not included in this literature review because the nature of the content is not related to the study of the sociocultural practice of brewing kombucha. That said, the motivations of San Martians to produce kombucha in their own homes can be situated within three circles of relevant literatures: a broader analysis of fermented foods and culture, fermented foods as part of an alternative food system, and the general relationships between community and belonging. Sandor Katz is largely considered the chief crusader for fermentation in the United States today. Katz believes practices of fermentation can be empowering and liberating and an avenue to engage with the world(s) around (and within) you. By becoming an active participator and producer, fermentos (which is Katz's term for fermentation enthusiasts) can reclaim their food systems and begin to improve their own health, as well as that of their communities and the world.

Fermented Foods and Culture

Fermentation is everywhere, yet the processes are rarely understood or appreciated. Several common foods are fermented, including cheese and bread, as well as coffee, chocolate, and vanilla beans (Tamang 2010). Fermentation has been a traditional technique of food preservation on all inhabited continents since Neolithic times (Soni and Day 2014). In Old World communities the baker and the brewer both played vital roles in turning the raw goods of grain crops into food products with a longer shelf life (beer) and a higher value (bread). Both of these production processes traditionally were accomplished with fermentation. As such, there is a broad body of literature on the history of fermented foods.

Microorganisms are ambient in the environment, and they transform food. There is a long and demonstrable history of symbiotic coevolution of humans and microbes (Sigal and Meyer 2016). Human species evolved with and within microbial environments, less out of human ingenuity and more out of the imperative and inescapable nature of microbes. In fact human survival is connected to microbes (Scott and Sullivan 2008). Since Pasteur's discovery of microorganisms, they have had a troubled reputation. However, when considering the sheer number of species that are not visible to human eyes, we must acknowledge that most microbes are "good" germs. Nevertheless, the predominant view that microbes are the source of disease has created a culture of sterilization wherein consumers make a concerted effort to remove bacteria from our bodies and our food (see also chapter 12). As a result, most of today's mass-produced and hypermarketed food is lacking in any life forms, even though humans evolved as a species in conjunction with a multitude of other life forms present in the foods and beverages we consumed (Abbot 2004).

Alternative Food Systems

Major aspects of this project lean heavily on the notion of an alternative food system; the alternative food system movement rejects the dominant food system and seeks a more regenerative, healthful, and locally oriented system of food production and distribution (Cleveland, Carruth, and Mazaroli 2015). From this perspective the current food system is one that has been manipulated by large corporations, with profit margins steering production methods. The history of the United States' food system is clear: since World War II Americans have been sold convenience foods—frozen, sterile, highly processed food products that lack nutritional value (Katz 2003). The alternative, championed by Sally Fallon (1999) and Michael Pollan (2006), among many others, advocates for "real" food instead: wholesome, nutritious food, traditionally cooked and prepared with intention and purpose. The alternative food movement is lauded as a solution to the ailments that come along with the American diet: health issues such as obesity, heart disease, diabetes, and cancer rates, all of which have all skyrocketed since the implementation of the modern food system. To combat these public health concerns, alternative food advocates suggest that we examine—and ultimately change—the system that produces these consequences.

Conventional farming practices and the current mode of food distribution have huge environmental impacts (Cleveland et al. 2016) and notable negative social externalities. The industrial agricultural model has many negative consequences for our planet, including soil degradation, excessive water use and waste, enormous petroleum-based energy use, and pollution from chemical fertilizers and pesticides to name a few (Belasco 2007). The agricultural industry is the number-one polluter and has a colossal impact on climate change and global warming.

Environmentalists tend to support agriculture reform (Cleveland et al. 2016) to counteract these atrocities.

Food activists are also fighting for social justice (Katz 2012). They argue that to revive local food production and local food exchange is to secure local food economics (Blake, Mellor, and Crane 2010). Social revolution is preceded by changing the means of food production (Katz 2006). By growing one's own food, people are seeking to reclaim their production of food and bring the scale down from a global perspective to a community-level one. The idea is that it is socially empowering to realize self-reliance in some form or another and that self-reliance is also good for environmental and personal health.

In this way local communities are strengthened when they support alternative food systems, which have taken many forms. Many people now engage in backyard gardening and keeping chickens, participating in community gardens, supporting local farmers by shopping at farmers' markets, and buying in to community-supported agriculture (CSA) programs. The homestead and do-it-yourself (DIY) movements are strong (Caruso 2015), indicating America's changing preference toward self-reliance and independence and away from the globalized agricultural industry. The focus on local and seasonal produce are the hallmarks of alternative food systems (Cleveland, Carruth, and Mazaroli 2015). Restructuring the production of our food can be the solution to so many of today's problems. Fermentos (the term used often by Sandor Katz) are active creators of unique products, as opposed to being passive consumers of mass-produced items. In sum, to reclaim one's food can be empowering, promote social and environmental justice, and contribute to healing our sick bodies and unhealthy world.

Community and Group Belonging

Food and community are two concepts that are intrinsically tied together. It's natural for food and people to be together, as we evolved as a species in small, collaborating communities. Community and culture are inherently enmeshed with food and the daily necessity of nourishment. Survival was ensured by collaborating as a group, including in the cultivation, preparation, and consumption of food. Sharing meals creates connections and opportunities for further group integration. Networks of sharing and exchange, mutual support, and caring are traced and reiterated through shared food practices.

As the prominent entomologist E. O. Wilson (2012) has noted in his work on superorganisms such as insect colonies and human societies, an individual's strong sense of belonging to the group is intrinsically tied to the success of that group. This innately human desire to belong to a group or tribe is telling of our evolutionary course. We evolved as a species in small groups or communities; our primate brains are still designed to manage the connections among about two hundred individuals, which is roughly the size of an ancient band of people. In today's disjointed and fragmented world, we are simultaneously and instantaneously connected to thousands of people, yet depression, anxiety, and loneliness plague our society, as we are isolated and removed from the group and community structure our brains crave. In other words, having a strong sense of community and connection to other the people leads to an increase in quality of life and to healthier, more productive members of the group.

Researchers have demonstrated how fermented food products and the making of these recipes can create tenacity and mindfulness (Santana et al. 2015). Fermenters fall into habits of cycle and are more in tune with seasonality and natural rhythms. The *Permaculture Book of Ferment and Human Nutrition* (Mollison 2011) discusses many of the methods of returning to traditional food preparation, production, cooking, and preservation techniques. Fermentation practices bring greater awareness to the interconnectedness of all species by highlighting the relationship between bacteria and yeasts, scoby and human (Scott and Sullivan 2008).

Moreover, fermented foods are unique, living expressions of place. The exact species of bacteria and fungi present in any symbiotic community will vary (Crum and LaGory 2016). Every scoby is a totally unique combination of species and will evolve to better adapt to its environment. Kombucha exemplifies the way food enables belonging and community building. Many cultural identities revolve around fermented foods. As previously stated, fermented foods have a long history with virtually every traditional culture (Kabak and Dobson 2011), and most cultures have developed their own unique ferments. Because a fermented food is basically in a state departed from fresh, but not yet rotten, the unique flavors can be an acquired taste. Persons belonging to a certain culture or ethnicity enjoy the taste and flavor of their particular ferment, but outsiders may find that food repulsive and disgusting (Pollan 2013). This effectively reinforces cultural identity through pride in culturally specific fermented foods.

Methods

This research project employed an ethnographic approach to explore and define kombucha culture.[2] Through cultural immersion and systematic investigation as both a participant and keen observer, the goal of this research was to gain an understanding of the motivations behind the practices that define this group of fermentos and to understand why they are brewing kombucha at home. To achieve this goal with integrity, I must let the group being studied speak for itself. Thus, I chose a combination of qualitative research approaches: semistructured interviews, focus groups, and participant observation to develop the rich, descriptive data produced by these methods. Together these techniques allow for a complete picture of the personal and collective experience of kombucha brewers to be accurately gleaned, and they illuminate key motivations that drive people to participate in this budding kombucha culture.

A total of twenty-five kombucha homebrewers participated in this project. Fifteen interviews were conducted, most of which were with individual brewers and a few with couples who brew together as a unit. In addition, two focus groups comprising seven participants each were also conducted to generate group consensus about why a kombucha practice was the common denominator of each group member. Most of the interview and focus group participants are actively involved in several local food and/or environment groups with which I am also engaged, such as the San Marcos Farmers' Market community and the group SMTX Permaculture, as well as the sister farms supported by that organization: Little Bluestem Farm, River Bottom Farms, Boxcar Farm and Garden, and Thigh High Gardens. In this overall community of groups I have been working alongside other kombucha brewers in planning and implementing meetings, workshops, and events. At one such event, while I was volunteering on the food prep team for the Hill Country Fair, a local festival of San Marcos culture benefiting local farms and gardens (Thigh High Gardens 2016), the conversation among the women chopping vegetables organically turned to fermentation and kombucha—and I paid close attention. This multifaceted, multimethod approach, using data from direct conversations about kombucha motivations and drawing on my own experiences within this community, allows for a thorough examination of what exactly constitutes kombucha culture in San Marcos, Texas.

Data Collection

Interviews

The fifteen different interview sessions were personal, one-on-one conversations in which kombucha brewers were asked to describe their experiences with kombucha and to indicate the motivation for their practice. A few brewers practice brewing kombucha with their partner, so I interviewed these couples together, as a unit. Because of my personal connection to so many brewers in the local community, reaching out to find kombucha enthusiasts to interview was not difficult. I first reached out to known kombucha brewers, inquiring by text, phone call, or in person, to see if they were interested in contributing to my project. I recruited additional participants through snowball sampling, asking interviewees for recommendations of people to whom they have passed along scobys or other people they knew of who were brewing kombucha at home. This sampling method achieved two things: first, it led me to other brewers to invite to participate and, second, it demonstrated the interconnected and interwoven nature of kombucha culture.

Keeping in mind that the main investigative focus of this study was the practice of brewing kombucha at home, I purposely conducted the interviews in the interviewees' own kitchens. The intention was to discuss kombucha brewing practice in the spaces where the brewer is most fundamentally engaged with kombucha culture. After the participants agreed to being interviewed at home, we arranged a mutually agreeable time to visit. Participants were given the alternative to have the interview take place at my own home or in a more neutral space like a coffee shop or the public library if for any reason they did not want to or could not accommodate the interview in their home. However, no one declined an in-home interview, as most brewers are quite enthusiastic about sharing their fervor for ferments in their spaces of fermentation.

The interview exchanges were informal and quite comfortable for all parties involved. I typically met with participants in the evening after both of us had experienced long workdays, but this did not diminish their enthusiasm for talking about kombucha. We chatted in their kitchens, as well as a few back porches and living rooms. As we talked, participants proudly showed off their scobys, some residing in a "scoby hotel"—essentially a controlled vessel for housing scobys not actively producing kombucha—and generously offered samples to me. In addition to kombucha, these kitchens were brimming with other food projects, such as the kraut fermenting in a beautiful aquamarine crock or the sourdough bubbling up in preparation for the next day's pizza dinner.

Participants told me about who first introduced them to brewing kombucha, revealing handwritten and dearly treasured instructions that were given to them as guides for the first few rounds and are now tucked away safely alongside other cherished recipes. Exact processes varied from brewer to brewer; some preferred to use both green and black teas, some were more experimental in the types of sugar utilized, and efforts to create unique flavor combinations were where the biggest differences occurred in practice. Similarities among many participants included being gifted their original scoby from a friend after discovering they had a taste for the fizzy beverage. A few participants told me about the specific health concerns that initially spurred their interest in fermented foods as an alternative path to healing.

These conversations gave a clear overview of how individuals practice brewing kombucha at home and why they initially got started brewing. These details were great starting points for understanding the connections within this group. Although full of informative and passionate expressions, the interviews did not immediately reveal deeper motivations for brewing or the strong sense of belonging that was gleaned once the group members were brought together. These semistructured conversations (aka focus groups) generated insight into the similarities and differences that individual group members brought to the collective.

Focus Groups

Fifteen people participated in focus groups for this study; most were also interviewed individually, but three people participated only in the focus group. As alluded to above, the focus group method is particularly appropriate for this study because the insights gained from dynamic group conversations facilitate sharing around the heart of the research question: What kinds of social bonds were created and/or enhanced through a shared practice of kombucha brewing? The group as a unit cooperatively generated knowledge, thus shifting the balance of power out of my role as the researcher and into the hands of the participants (Bosco and Herman 2010). I met on two occasions with groups of kombucha brewers at different local farms. One occasion was after an SMTX Permaculture meeting at Thigh High Gardens. As the brewers from the group gathered to discuss their kombucha practices, a few other permaculture enthusiasts lingered around the open-air patio. As the night encroached around us and the group conversation explored the connections we had forged through brewing, the others' interest in brewing their own kombucha grew. By the end of the evening they were rewarded with their own scobys to take home, along with some personalized instructions on how to begin the process.

The other communal meeting was held on a different local farm, Boxcar Farm and Garden. A group of seven participants gathered at a large handcrafted dinner table and collectively chopped vegetables in preparation for serving food at an upcoming community festival. The group conversation that night was lively, with colorful stories exchanged about everyone's first introduction to kombucha. Both discussions were generally democratic and not necessarily dominated by any particularly vocal participant, although some members tended to stay quiet until directly addressed. By steering the discussions toward deeper topics, such as the connections between members and the strong sense of community among members of the group, we all explored together how impactful kombucha brewing had been in forming social bonds between members. Even though this acknowledgment came from within the group itself, I think it was a surprising revelation to some and might not have emerged without the vibrant discussion of the focus group. The conversations generated from the supportive, collaborative setting of the focus group led directly to a better understanding of the networks, bonds, sense of community, and togetherness that this group experiences.

Participant Observation

Participant observation in the context of this project involved describing my own experiences of being a member of this community group and providing detailed descriptions and thoughtful reflections after interactions with others and following my own individual brewing sessions. This research topic was organically generated from my own unique position and experiences. As such, in the tradition of the feminist researcher Gillian Rose (1997), I carefully examined my own position within this group of fermenters and acknowledge that my interactions with the group helped to shape the data generated, thus influencing the knowledge that was produced from these interviews and focus groups. This project came to fruition as two parts of my life came together: completing graduate study in geography, including studying critical qualitative methods, and pursuing my budding passion for brewing kombucha at home with the support of a larger group of fermenters. The merging of these two aspects of my life has resulted in a unique project that could not likely have been conceived of by anyone else.

I have been immersed in this mutual community of kombucha culture since before this project was even an inkling of a thought. As an active member of SMTX Permaculture since 2013, I volunteer on local farms and help with other events promoting sustainable agriculture, namely the Hill Country Fair. Another event at which brewers convened was the Austin Fermentation Festival in October 2016. One of the interviewees who is particularly committed to kombucha, among other many other ferments and home food projects, led a hands-on fermentation class and workshop at Little Bluestem Farm that I attended in the spring of 2016. By being active and engaged with brewers in so many other venues besides the formal interviews and focus groups, I could gain a deeper and broader understanding of why people are so committed to the product and practice of kombucha.

Analysis Techniques

I wanted to be able to participate in and facilitate the conversations with brewers, so it was important for me to record them so as to dispense with the distraction of writing down responses. Accordingly, the interviews and focus groups were digitally recorded on my iPhone using the voice memos app. In addition, after some of the interviews I took photographs of the brewers in the kitchen as they proudly displayed their food projects brewing on the counter. I took detailed field notes, specifying the conditions before the interview and reflecting on our conversations afterward, referring to the voice recordings often. I chose not to transcribe the audio recordings for many reasons, chiefly lack of money and time, but also because the scope of this project does not necessitate doing so. After attending events, I would create a field note entry describing the experience and any observations I made.

After collecting the data, I analyzed them in myriad ways. I began by combing through my interviews, notes, photographs, and recordings, pulling out what could be evidence of motivation to participate in kombucha culture. Three broad categories evolved, and I then used focused coding techniques to further scrutinize the data. Coding the data served to focus and condense the themes, enabling me to identify patterns, understand connections, and elicit meaning from the observations and interviews. From this analysis, three main motivations for homebrewing kombucha were revealed, each of which will be presented later in this chapter.

After I had generated the three themes, I went back to some of my original interviewees and presented my interpretations. This qualitative analysis technique is known as member-checking, wherein the research findings and conclusions are shared with the informants to see if the results reflect what they intended. This is a key step in shifting power from the researcher to the research subjects (Watson and Till 2010). Through member-checking, I successfully sought to increase the validity and rigor of the research: the brewers all concurred with my findings that three main reasons motivated a kombucha homebrew practice.

Findings and Discussion

Thoughtful analysis and consideration of data generated from the interviews, focus groups, and participant observation illuminated three key themes that explain why brewers are motivated to begin and maintain their kombucha practice. The first theme revolves around the health and wellness benefits one receives from regular consumption of the beverage and the healthy habit of homebrewing. The second reason people are motivated to make kombucha at home is because they view their kombucha practice as part of their active engagement with alternative food systems. The third theme that emerged from the data is that homebrewing kombucha is part of a process of cultivating a sense of community. The combination of these common motivations, communal beliefs, and shared practices culminates in what I term "kombucha culture." The group of homebrewers in San Marcos described herein demonstrates kombucha culture through many different avenues of expression. The presence of kombucha culture here is illustrative of a greater phenomenon happening in other locales around the nation. It is insightful to consider why so many people have a practice of brewing kombucha at home.

Health and Wellness

When asked what motivates their practice, most brewers began with their concerns for personal health and well-being. Kombucha has a reputation as a healthful product, and the pursuit of better health was the primary reason that most of the participants ventured into the world of ferments—and specifically into kombucha. Kombucha is touted as a functional food, with nutritional benefits that go beyond providing basic nutrients; these additional benefits include vitamins, minerals, antioxidants (proven to possess cancer-fighting/preventative qualities), and amino acids. One respondent, Alex, recalled that her first encounter with kombucha was at a small health foods shop in Austin where she was introduced to the beverage by a group of her "crazy old-school Austin hippie friends who were really into macrobiotic food."

One of the biggest reasons people brew and consume kombucha is for the live cultures of yeast and bacteria that can help restore gut microbiota. The benefits of eating live foods for gut health is an emerging field, and the importance of a healthy digestive system to overall health is becoming more apparent (Chaves-López et al. 2014). Improving the intestinal tract can lead to better digestion. One respondent, during the first focus group, recounted the struggles she underwent with her digestive system, including unpleasant experiences with IBS. Her health issues were really starting to negatively affect her life, so she decided to take action and combat her symptoms through a diet that focused on digestive health. This led her down the path of ferments—and eventually to kombucha. I recall she was the first of my friends to lend me one of Sandor Katz's books, *Wild Fermentation*.

Another benefit from regular kombucha consumption is a strengthened immune system. Scientists are just beginning to understand the connection between gut health and psychological health, but the implications are significant. Some researchers believe the root cause of conditions such as ADHD/ADD and anxiety is poor gut health (Campbell-McBride 2010). Consuming a diversity of live foods promotes a healthy, diverse gut flora that in turn can encourage good mental health.

Kombucha consumption can also provide immunity support. Ninety percent of our physical immune system is located in the walls of our intestines; bolstering gut health in turn supports top immune system function (Hooper, Littman, and Macpherson 2012). There is anecdotal evidence that kombucha can even help prevent cancer because of its antioxidant properties. Katz's original engagement with fermentation was in response to a diagnosis of HIV. Katz, seeking alternatives to Western medicine, found a group of people living on a small farm in Tennessee practicing food traditions that incorporated a whole array of ferments across the diet in support of living a healthy life despite HIV/AIDS. He has incorporated ferments of all kinds into his diet for twenty years and has seen a marked improvement in overall health.

One interviewee reported an unusual scenario. Her twin sister suffered a very serious rattlesnake bite to the foot in early spring 2014. In the five days she spent in the hospital she was administered twenty-five vials of antivenin (hospital doctors told her a typical dosage was only four). This particularly severe snakebite, coupled with a preexisting weak immune system, left her immobile for months. As her body processed the extreme amount of neutralized venom, her mind recovered from trauma. She incorporated the regular consumption of kombucha into her diet, believing it would help improve her overall level of health and wellness.

Another unexpected response that was reported multiple times was the consumption of homebrewed kombucha as a healthier carbonated beverage alternative to soda or beer. Brewer Judith talked about her motivation to brew kombucha at home so her partner would drink less Dr Pepper, a carbonated soft drink loaded with high-fructose corn syrup.

I have heard many testimonies about a serious addiction to soda that was curbed or eliminated by turning instead to kombucha. Kombucha is also an option for those who opt not to drink alcoholic beverages for any number of reasons. Lynn mentioned that he enjoys having kombucha in place of beer when he wants to make a healthful choice but still enjoy a cold, refreshing, bubbly drink: "It's nice to have a drink to share and enjoy with others while socializing without having to ingest alcohol."

Beyond the health benefits, one of the most common explanations for maintaining a kombucha practice was simply because "it tastes good!" as stated by Veronica. A lightly effervescent, flavorful beverage is enjoyable and refreshing, especially in the sweltering summer heat. Because fully brewed kombucha isn't chock full of sugar and caffeine, it is an enjoyable beverage to consume on a hot day. In addition, the practice of brewing kombucha, waiting for it to ferment, and then continuing to bottle and brew again is a cultivated habit. Staying in tune with your scoby and the kombucha fermentation process keeps you aware of more natural cycles and rhythms.

Engagement with Alternative Food Systems

Fourteen out of eighteen respondents referred to feeling engaged with alternative food systems by brewing kombucha at home. Kombucha is one of the most easily accessible and least intimidating fermentation processes, so it's no surprise that many fermentos begin their journey into the world of ferments with kombucha. It has relatively low startup costs and minimal inputs, especially compared to more complex fermentation processes, such as beer homebrewing. Plus, bottled kombucha sold in stores is much more expensive than kombucha brewed at home. In addition, homemade food and beverage products decrease dependency on supermarkets and the agroindustrial food system, which degrade our environment and squander resources in the name of profit. Practices such as brewing kombucha at home aim to mitigate some of the negative environmental and ecological impacts of mass food production and distribution. In fact almost every participant in this project was directly engaged with a local, alternative food system: some brewers are backyard gardeners, some run small holistic farms, some practice permaculture and restorative agriculture techniques. A commitment to the availability of high-quality, locally grown, seasonally appropriate produce was a common thread among participants.

Alternative food systems support local community markets through noncash exchanges. Homemade kombucha can be gifted or traded for other valuable goods. Multiple interviewees reported using kombucha to barter for other goods, such as produce or canned items. Ali even reported that she was compensated in gallons of kombucha for volunteering at a local farm. Many also named gifting kombucha as one way they engage in alternative economies. As this evidence suggests, kombucha brewing creates empowering change on three nested scales, allowing homebrewers to reclaim their food and begin to improve the health of individual bodies, wider communities, and the planet as a whole.

Sense of Community

The third theme that this study brought to light is the focus on the communal aspect of brewing kombucha at home. Almost every single brewer involved in this project mentioned this theme in one way or another. When asked about how they originally became involved in homebrewing kombucha, most respondents said they were introduced to ferments by close friends. The previous themes of health and wellness as well as engagement with alternative food systems are important, but I believe this final theme to be the most important reason that people are brewing kombucha: its critical contribution to the development of kombucha culture. How exactly does homebrewing kombucha cultivate a sense of community? The data show that it does so through networks of homebrewers sharing their passion, through friendships and bonds forged through the practice, and from the collective sense of pride that kombucha brewers possess for both their homemade elixirs and the hometown that gave rise to this phenomenon.

Social networks (not of the digital variety) are created as one brewer

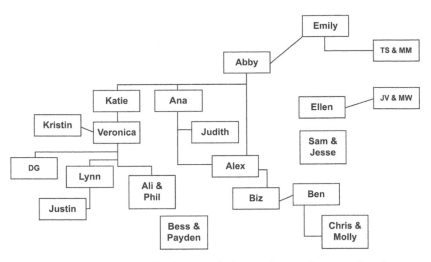

FIG. 37. Example of a scoby family tree, which maps how scobys were shared among homebrewers. Created by Elizabeth Yarbrough, 2017.

passes to a friend a scoby and instructions on how to brew. For example, interviewee Veronica talked about first acquiring a scoby from her friend Katie: "Katie gave me a scoby, but I didn't really know what to do with it. So I went over to her house and watched her brew. I took lots of notes, as you know I do, and Katie was really patient as I asked a lot of questions." The exchange of the physical materials such as the scoby, starter kombucha, and maybe even a brewing vessel is enhanced by the exchange of information about the brewing process. Through sharing ideas and materials, webs of brewers emerge. The scoby family tree (fig. 37) is a flow chart depicting the exchange of scobys between homebrewers in San Marcos and created with data from the interviews and focus groups.

As you can see from the chart, Veronica acquired her scoby from Katie, who acquired her scoby from Abby. Veronica also gave away iterations of her scoby to Kristin, DG, Lynn, and Ali and Phil. Some of the brewers included in this study did not receive their scoby from others involved in the project. These brewers (Bess and Payden, Sam and Jesse, Ellen, etc.) acquired their scobys either from other friends, online mail order, or growing their own from a plain bottled kombucha from the store. The chart demonstrates how the practice of homebrewing can spread and creates a web connecting people through the common thread of kombucha.

This study revealed some other major elements of kombucha culture, including the importance of the scoby and the wider networks built through its exchange. For example, by swapping the physical material of the scoby, fermenters create exchange networks and establish mutually beneficial friendships based on sharing and common interest. Trust develops out of these interactions, and fermenters then feel as though they can rely on each other for trading other resources, such as tools and supplies, information and advice for troubleshooting problems, or for feedback or help in implementing new ideas. These networks are strengthened as social connections become deeper and last longer. The significance of this factor is apparent when you consider that the opportunity for personal networking and sharing of genetic material heritage is lost when scobys are purchased commercially. Out of the homebrewers to whom I spoke, only two people utilized these more detached methods, and that was simply because at the time they began brewing, they didn't know anyone with a scoby in active production. Both of these participants expressed appreciation and respect for the human-to-human exchange process and have since shared their scobys with other new brewers.

Brewers providing feedback and support concerning fermentation issues also tend to share concern and advice for each other's life problems. Interactions between individuals, such as scoby swapping and shared practices such as brewing together, create bonds and cultivate friendships. These friendships are reinforced through the sharing of the final brew product. One of the more unexpected findings of this study was how many people talked about sharing their brews with friends while entertaining. During my interview with Lynn, he talked about how much he loves to "bust out the 'buch'" when friends gather at his home or to bring it to parties. He expressed a real sense of pride in his creation and a need to share this with his friends. This sharing factor ties into a sense of community because the members want to give back to the group.

Another way these networks are reinforced is community events. For example, most of the brewers involved in this project attended the third annual Hill Country Fair in November 2016, hosted at Thigh High Gardens (Thigh High Gardens n.d.). At this festival homemade fermented beverages were spotlighted, and there were multiple varieties of kombucha available for purchase from different vendors. Other events, such as a fermentation class that interviewee Abby held at Little Bluestem Farm, brought together individuals who shared a common interest in learning more about home fermentation projects, including kombucha. Some respondents admitted that it was less the subject matter and more the classmates that brought them to a fermentation class, thus highlighting the importance of a sense of community. Interviewee Kristen told me (referencing a different social occasion), "Anything I can learn with a group of girls is awesome!"; clearly, communal learning can be an effective tool in strengthening groups. Through many other shared community events less directly related to kombucha, such as farm work days and organizing friends' weddings or even memorial services, this network of brewers grows and a sense of community is enhanced.

San Martians have an enthusiasm for their community, as demonstrated when Ali declared "these are my people!" when I inquired about her connections to other brewers in town. A collective sense of pride for San Marcos as their hometown is shared among the brewers, even though not all of them currently live in the vicinity. This love of place unites brewers and gives the network a background or stage. This network of brewers produces friendships and belonging, which in turn evolve into group identity and belonging—chief components of a strong sense of community.

Conclusions

This chapter has explored a particular kind of "fermented landscape," one revolving around the cultures of kombucha and cultural practices of kombucha fermentos. As the term "culture" connotes multiple definitions from anthropology, biology, and beyond, throughout this project I have utilized the term "kombucha culture" to evoke a blended meaning of the word *culture* in reference to the group of people who create networks of shared ideals, beliefs, and material practices related to brewing kombucha. The scoby family tree shows the connections between fermenters—a physical tracing of the spreading of culture, specifically a kombucha culture. Kombucha culture satisfies an innately human need to be included and accepted by the group (Wilson 2012). Through sharing and exchanging information and materials, brewers become more connected through social networks. Friendships are created and strengthened through the practice of brewing kombucha. By sharing common goals, as well as a collective sense of pride for kombucha products, brewers build a collective identity.

In response, this project's goal was to understand why so many of my community members were passionate about brewing kombucha at home. As demonstrated through the dialogue generated during interviews, the conversations of focus groups, and my own immersive experience with this group of hometown fermentation enthusiasts, three key pillars of kombucha culture emerged: first, a concern for personal health and well-being; second, an active engagement with at least some aspects of alternative food systems; and third, a commitment to a strong and vibrant local sense of community.

While this project was small in scale and locally based, the project findings can be applied to much more than homebrewing kombucha in San Marcos. By identifying the particular motivations of local brewers to engage in kombucha culture, we simultaneously have defined values with universal appeal. These three themes echo the sentiments of many people across the country, and indeed around the globe, who are striving for a better life in today's uncertain world. Kombucha culture is the expression of a collective desire for a more healthful and balanced lifestyle in a supportive community striving for a better—local, seasonal, organic—food system. Thus, the insights garnered from this study resonate much more broadly.

Notes

1. Throughout this chapter, the personal pronoun "I" refers to the chapter's first author, Elizabeth Yarbrough.
2. Careful consideration was taken to assure the safety and security of all respondents involved in this project. Because of my close familiarity and the generally benign nature of this project, there was very little risk of harm resulting from their participation. An application for an exemption of review was submitted to Texas State University's institutional review board (IBR) and approved for an exemption of full review.

References

Abbott, Alison. 2004. "Microbiology: Gut Reaction." *Nature* 427 (6972): 284–86.

Belasco, Warren J. 2007. *Appetite for Change: How the Counterculture Took on the Food Industry*. Ithaca NY: Cornell University Press.

Blake, Megan K., Jody Mellor, and Lucy Crane. 2010. "Buying Local Food: Shopping Practices, Place, and Consumption Networks in Defining Food as 'Local.'" *Annals of the Association of American Geographers* 100 (2): 409–26.

Bosco, Fernando. J., and Thomas. Herman. 2010. "Focus Groups as Collaborative Research Performances." *The Sage Handbook of Qualitative Geography*, edited by Dydia DeLyser, Steve Herbert, Stuart Aitken, Mike Crang, and Linda McDowell, 193–208. London: SAGE.

Campbell-McBride, Natasha. 2010. *Gut and Psychology Syndrome: Natural Treatment for Autism, Dyslexia, Depression, Dyspraxia, A.D.D., A.D.H.D., and Schizophrenia.* Rev. ed. Cambridge: Medinform Publishing.

Carr, Coeli. 2014. "Kombucha Cha-Ching: A Probiotic Tea Fizzes Up Strong Growth." CNBC, August 9, 2014. http://www.cnbc.com/2014/08/08/kombucha-cha-ching -a-probiotic-tea-fizzes-up-strong-growth.html#.

Caruso, Christine C. 2015. "Home Fermentation." In *The SAGE Encyclopedia of Food Issues*, edited by Ken Albala, 2:792–94. 3 vols. Thousand Oaks CA: SAGE.

Chaves-López, Clemencia, Annalisa Serio, Carlos David Grande-Tovar, Raul Cuervo-Mulet, Johannes Delgado-Ospina, and Antonella Paparella. 2014. "Traditional Fermented Foods and Beverages from a Microbiological and Nutritional Perspective: The Colombian Heritage." *Comprehensive Reviews in Food Science and Food Safety* 13 (5): 1031–48.

Cleveland, David A., Allison Carruth, and Daniella N. Mazaroli. 2015. "Operationalizing Local Food: Goals, Actions, and Indicators for Alternative Food Systems." *Agriculture and Human Values* 32 (2): 281–97.

Cleveland, David A., Lauren Copeland, Garrett Glasgow, Michael Vincent McGinnis, and Eric R. A. N. Smith. 2016. "Influence of Environmentalism on Attitudes toward Local Agriculture and Urban Expansion." *Society & Natural Resources* 29 (1): 88–103.

Crum, Hannah, and Alex LaGory. 2016. *The Big Book of Kombucha: Brewing, Flavoring, and Enjoying the Health Benefits of Fermented Tea*. North Adams MA: Storey.

Dixon, Richard, and Stacy Bray. 2010. *Appendix P4 Preliminary Texas Evaporation Trends Report*. https://gato-docs.its.txstate.edu/jcr:d97dab4e-67fe-45e5-8303-1cf064adcfca/Preliminary_Texas_Evaporation_Trends_Report.pdf.

Fallon, Sally. 1999. *Nourishing Traditions: The Cookbook That Challenges Politically Correct Nutrition and the Diet Dictocrats*. Brandywine MD: New Trends.

Frank, Günther. W. 1995. *Kombucha: Healthy Beverage and Natural Remedy from the Far East*. Steyr, Austria: W. Ennsthaler.

Hobbs, Christopher. 1995. Kombucha, Manchurian Tea Mushroom: The Essential Guide. Santa Cruz CA: Botanica Press.

Hooper, Lora. V., Dan. R. Littman, and Andrew. J. Macpherson. 2012. "Interactions between the Microbiota and the Immune System." *Science* 336 (6086): 1268–73.

Kabak, Bulent, and Alan D. W. Dobson. 2011. "An Introduction to the Traditional Fermented Foods and Beverages of Turkey." Critical Reviews in Food Science and Nutrition 51 (3): 248–60.

Katz, Sandor Ellix. 2003. *Wild Fermentation: The Flavor, Nutrition, and Craft of Live-Culture Foods*. White River Junction VT: Chelsea Green.

———. 2006. *The Revolution Will Not Be Microwaved: Inside America's Underground Food Movements*. White River Junction VT: Chelsea Green.

———. 2012. *The Art of Fermentation: An In-Depth Exploration of Essential Concepts and Processes from Around the World*. White River Junction VT: Chelsea Green.

Kaufmann, Klaus. 2013. *Kombucha Rediscovered! The Medicinal Benefits of an Ancient Healing Tea*. Summertown TN: Books Alive.

Marsh, Alan J., Orla O'Sullivan, Colin Ray Hill, R. Paul Ross, and Paul D. Cotter. 2014. "Sequence-Based Analysis of the Bacterial and Fungal Compositions of Multiple Kombucha (Tea Fungus) Samples." Food Microbiology 38:171–78.

Mollison, Bill. 2011. The Permaculture Book of Ferment and Human Nutrition. Tasmania, Australia: Tagari.

Office of Institutional Research. 2017. Facts and Data: University Marketing. Texas State University. Accessed May 21, 2019. https://brand.txstate.edu/facts-and-data.html.

Paxson, Heather. 2008. "Post-Pasteurian Cultures: The Microbiopolitics of Raw-Milk Cheese in the United States." *Cultural Anthropology* 23 (1): 15–47.

Pollan, Michael. 2006. *The Omnivore's Dilemma: A Natural History of Four Meals*. New York: Penguin Press.

———. 2013. *Cooked: A Natural History of Transformation*. New York: Penguin Press.

PR Newswire. 2016. "Kombucha Market Worth USD 1.8 Billion USD [*sic*] by 2020." PR Newswire Association. Accessed December 15, 2016. http://www.prnewswire.co.in /news-releases/kombucha-market-worth-usd-18-billion-usd-by-2020-522394761.html.

Pryor, Betsy, and Sanford Holst. 1996. *Kombucha Phenomenon: The Miracle Health Tea; How to Safely Make and Use Kombucha*. Sherman Oaks CA: Sierra Sunrise.

Rose, Gillian. 1997. "Situated Knowledges: Positionality, Reflexivities and Other Tactics." *Progress in Human Geography* 21 (3): 305–20.

San Marcos Greenbelt Alliance. 2015. San Marcos Natural Areas. http://smgreenbelt .org/natural-areas/.

Santana, Christina, Stacy Kuznetsov, Sheri Schmeckpeper, Linda J. Curry, Elenore Long, Lauren Davis, Heidi Koerner, and Kimberly Butterfield McQuarrie. 2015. "Mindful Persistence: Literacies for Taking Up and Sustaining Fermented-Food Projects." *Community Literacy Journal* 10 (1): 40–58.

Scott, Robert, and William C. Sullivan. 2008. "Ecology of Fermented Foods." *Human Ecology Review* 15 (1): 25–31.

Shiner, Joel L. 1983. "Large Springs and Early American Indians." *Plains Anthropologist* 28 (99): 1–7.

Sigal, Michael, and Thomas F. Meyer. 2016 "Coevolution between the Human Microbiota and the Epithelial Immune System." *Digestive Diseases* 34 (3): 190–93.

Soni, Surabhi, and Gargi Dey. 2014. "Perspectives on Global Fermented Foods." *British Food Journal* 116 (11): 1767–87.

Stamens, P. 2005. *Mycelium Running: How Mushrooms Can Help Save the World*. Berkeley CA: Ten Speed Press.

Tamang, Jyoti Prakash. 2010. *Fermented Foods and Beverages of the World*. Boca Raton FL: CRC Press.

Thigh High Gardens. n.d. "Hill Country Fair." Accessed May 21, 2019. https://www .thighhighgardens.org/fair-info.

U.S. Census Bureau. 2015. "Ten U.S. Cities Now Have 1 Million People or More; California and Texas Each Have Three of These Places." https://www.census.gov/newsroom /press-releases/2015/cb15-89.html.

Watson, Annette, and Karen E. Till. 2010. "Ethnography and Participant Observation." In *The SAGE Handbook of Qualitative Geography*, edited by Dydia DeLyser, Steve Herbert, Stuart Aitken, Mike Crang, and Linda McDowell, 121–37. Los Angeles: SAGE.

Wilson, Edward O. 2012. *The Social Conquest of Earth*. New York: Liveright.

Part 3

Perspectives on the Possibilities and
Limitations of Linking Fermentation
and Landscape

Fermentation and Kitchen/ Laboratory Spaces

10

Maya Hey

When I worked as an apprentice at a Michelin-star restaurant, I found myself picking thyme leaves with tweezers and scrubbing stainless steel countertops every eight hours. I also plated desserts using liquid nitrogen, a tank of which sat unassumingly in the corner of the prep kitchen. During my chemistry days, this liquid nitrogen was a hands-off matter, yet at the restaurant I was refilling kettles of liquid nitrogen on a nightly basis. Although the material ingredient was the same, liquid nitrogen was understood differently and thus handled differently in these two spaces. The same material in two different spaces demonstrates how processes of knowledge production are subject to the different sets of values, epistemics, and regulations that animate each space. A similar encounter—with milk—forms the basis of this chapter, in which other landscapes are considered to study fermentation's ability to transform materials and produce meanings.

This chapter examines the archetypes of commercial kitchens and scientific laboratories as spaces for (food) knowledge production. How do these spaces contribute to the formation, exchange, and metabolism of (food) knowledge? Specifically, how do the bodies (that is, human bodies, microbial bodies, regulatory bodies) move in and through these spaces? These spaces are not just passive backdrops; they affect how certain kinds of knowledge become credible, accepted, and circulated over others. Their material surroundings (such as stainless steel countertops) and the ritualized practices born of these environments (such as sanitation protocols) reinforce an ideology of safety that, while helpful in certain instances, can perpetuate the myth of human mastery. Comparing these spaces as arenas for food research can open up questions about exactly who or what is contributing to processes of knowledge production. Since humans are not the only ones who ferment in these landscapes, we can begin to reflect on other forms of agency that we may often take for granted.

In fermentation, human agency works alongside more-than-human forces in ways that disrupt our understanding of causality. Fermentation is a process of transformation due to microbes like bacteria, molds, and yeasts that transform food ingredients both materially and symbolically; the resulting ferment is just as much a product of time and ambient factors as it is a product of interactions between humans and microbes.[1] As a result, decentering the human is necessary to better account for a more holistic and complex understanding of how food and knowledge are produced in spaces like kitchens and laboratories.

Studying these spaces can highlight processes of power (institutionally imposed, socially constructed) that shape the meaning of food. Kitchen and laboratory spaces valuate microbial life differently, including which microbes count as safe or unsafe and how to regulate them when their populations become unruly. Since fermentation necessitates an encounter with microbial life, these spaces directly affect the formation, exchange, and metabolism of food knowledge. For example, in analyzing Louis Pasteur's early microbiology work, Bruno Latour (1988) highlights the fact that Pasteur's discoveries were made in the context of microbial disease and decay, thereby solidifying the microbe's relation to humans only in terms of harm (see chapter 12 for a related discussion). The ontological fixedness of a microbe (e.g., their pathogenicity) is emergent, not anchored, raising questions about how we relate to and work with microbial life in food settings. By calling attention to the material actors and the immaterial factors that move through these spaces, we can recognize processes of power as they take shape.

This chapter presents a case study that complicates commonly held ideas about food, including safety and edibility. Using the methodology of research-creation, a food provocation is staged for a public audience. Research-creation is a practice-based research methodology defined to be an "approach to research that combines creative and academic research practices . . . through artistic expression, scholarly investigation, and experimentation" (SSHRC 2016). Owen Chapman and Kim Sawchuk (2012) use Wittgenstein's conceptual framework of family resemblance to enumerate four iterations of research-creation. One of these, "creation-as-research," acknowledges that "one way 'to know' is 'to do'" (14) and prioritizes process over outcome (21). Natalie Loveless (2015, 42) argues that research-creation "means embracing an interdisciplinarity with regard to practice/theory lines"; elsewhere she has described research-creation as a neologism of "working practicetheoretically" (2012, 93). When studying fermentation, research-creation allows for a situated analysis of the material practice alongside theory so that these two epistemologies can inform one another.

In the case study presented, milk is transformed into an experiment that challenged conventional notions of edibility. Milk is intentionally inoculated with microbes taken from a hand swab in order to make cheese, complicating the boundaries of body, microbe, and food. Working with the same material in two different spaces (milk in kitchens versus milk in laboratories), as well as the same material reality at two different sites (bacteria on hands versus bacteria in foods), shows that there are more materials (and discursive relations) at play than what is normally considered actionable.

Using the material-semiotic analysis of this milk and the cheese produced from it, this chapter examines other landscapes that can produce meanings. It attempts to locate agency in human and more-than-human loci so that we can begin to unravel the ideological narratives that dominate terrains of food knowledge. First, a conceptual comparison of archetypal kitchen and laboratory spaces will elucidate how safety and control are constructed in these spaces. Then, in the form of a case study, this chapter will use insights from a particular food provocation to consider a more layered understanding of safety and contamination, with a particular eye toward gaining a better understanding of how we live with microbial life. A theoretical discussion will follow and will propose a shift in ontological perspective to better understand how our view of microbes informs larger scales of society.[2]

The Theory and Practice of Aspirational Control and Safety

Microbes live in, on, and around us, even though we cannot easily see them. Because of their ubiquity, commercial kitchens and scientific laboratories have evolved to become controlled spaces such that work surfaces are under constant microbial watch. These spaces are sterile and frequently resterilized to maintain control.[3] One manifestation of this control is in how the human instigators (that is, chefs or scientists) frequently and ritualistically clean their space. Revisiting the vignette at the Michelin-star restaurant, I can recall washing all of the work surfaces, the sidewalls, the floors, and the inside of overhead vents twice a day, every day. Food-grade chemical sanitizers like quaternary ammonia were always within reach, just like the spray bottles of denatured ethanol that dotted the workbenches of the biology labs I used to frequent. Both spaces were furnished with stainless steel counters, the standard for nonporous surfaces, so that contaminants cannot hide and taken refuge in the microscopic pores of ceramic, cloth, or wood.

These spaces locate the human instigator at the center of scientific and culinary investigation, keeping ambient contaminants to a minimum. Contaminants in both spaces could be defined as unwanted or unintended species in spaces demarcated as safe (to humans).[4] Both obliterate nonhuman traces with heavy-handed use of cleaning agents; otherwise, the assumption is that contaminants could run amok, invisible to the naked eye, until their manifestations become visible and it is too late to contain them.

The material surroundings of kitchen and laboratory spaces frame the human instigator as principal (as in principal investigator) or chief (as in *chef du partie*): lab coats and chef's whites protect and signify expertise, extraction vents and exhaust fans control air flow (steering air away from the instigator), and everything is labeled in permanent marker to assure clear communication between instigators. It appears that each of these materials serves a protective function to keep instigators at a safe distance from what could potentially compromise or contaminate them. These materials animate the ritualized practices of the space (like hand-washing and glove-changing) and shape the instigators' perception of safety. The repetition of these performances constructs a syllogism whereby control equates with safety. While such measures could be understandable, the perceived "better safe than sorry" mentality makes overprotection the norm.

Underlying this paradigm is the assumption that human beings and microbial life are distinct entities that interact; in fact we are quite enmeshed.[5] It becomes clear that these spaces are made for the human user, who harnesses and utilizes microbes as a functional tool. The separation of human and microbe as distinct entities permits the ideological justification of capitalizing on microbial labor. Contemporary kitchens now boast that they have fermentation programs, make in-house seasonings, or reclaim would-be food scraps (e.g., the fibrous base of celery stalks) to transform them into fermented delights (e.g., celery vinegar). Laboratories often use *E. coli* or yeasts as "model" species to study genomic or enzymatic activity in living systems because their significantly shorter life cycles permit the collection of multigenerational data within a short period of time. In each space the first steps of protocols and recipes include establishing a "clean slate" to favor the growth of the target microbe, such that the instrumentalized microbe lives or dies according to the instigators' will and ethics. Although these safety procedures are meant to protect us from *particular* microbes, institutional pathways for operationalizing this safety attempt to eliminate *all* microbes. However aspirational, this rhetoric of control follows the Pasteurian logic of eradicating all microbes and specifically creating thereafter environmental conditions in which only the intended microbe can thrive.

Thus, it is in the context of use and utility that humans and microbes interact in kitchen and laboratory spaces. This ideology sets up one of two power relations, both of which privilege the human instigator: antagonism or exploitation. Even the post-Pasteurian logic that Heather Paxson (2008) outlines in her analysis of raw-milk cheese seems to emphasize the microbe's potential as a probiotic, not immediately for the human end-user but to enhance the ripening of the cheese (which is ultimately destined for human consumption). Utility requires an ontological separation (of us/them, for example) if only to identify a "them" to mobilize. Here the spatial separation between worker and worked-on (i.e., between subject and object, between human and microbe) becomes the basis for determining what is to be othered. In turn, such practices reify our sense of purity and primacy, which contributes to the myth of human exceptionalism. This ideology fuels the kind of categorical elimination versus a more nuanced approach in which we draw lines of permissibility (which comes with its own set of problematics, including broad-spectrum antibiotics that eliminate gut flora or the alarmingly high rates of antibiotic resistance). Perhaps our steadfast belief in control and safety is too narrowly defined in that the normalization of (and subsequent expectation for) human power is seeing its limits. How could processes of fermentation drive social and theoretical change in redefining human positionality? How do we make these ethical decisions, especially at larger scales of public function and commensality?

Case Study of a "Handmade Cheese"

In 2016 I staged a food provocation to open up conversations around what constitutes contamination, or how certain foods were made to be unsafe. In a "handmade cheese" I replicated the experiments of the biologist Christina Agapakis and made a cheese using milk that had been incubated with the bacteria living on my hands and the hands of three other university staff members.[6] Rather than remain in the conceptual realm of how we relate to microbes, I made this cheese because its physical form raises questions about what goes into our notion of edibility, what constitutes contamination, and how we decide to eat other living beings. I wanted people to eat a "handmade cheese" to imbricate their physical, corpus-based ethics with philosophical gymnastics: if microbes make up part of the human biome, then there is a literal piece of human in this cheese *and every other food ever made.* How does knowing this affect one's ideas of handmade food writ large? By overemphasizing the bacterial nature of an otherwise edible form of the cheese, I was asking regulatory bodies and public bodies to reconsider the lines of permissibility. Eating would only enrich this pending conversation.

The Rationale

The first step of cheesemaking requires acidifying milk, which helps the casein proteins coagulate later (when the cheese curds solidify). Normally this acid comes from adding lemon juice, splashing in some buttermilk, or letting the milk "sour" overnight. As a substitute, bacteria were swabbed from a human hand and added to milk so that they would digest lactose (or milk sugar) and produce lactic acid. The resulting cheese follows the know-how of cheesemaking as part of a much deeper narrative between humans and microbes. For millennia ferments and their microbial cultures helped preserve foods that would otherwise spoil without refrigeration. Intentionally inoculating milk with an array of bacteria and molds would ward off opportunistic microbes in the ambient environment, keep the milk from spoiling, and prolong the milk's edibility over weeks, months, or even years. Microbes have provided a critical pathway for food security in which perishability was negotiated, not fixed. The "handmade cheese" was both a testament to the long legacy of human-microbe relationships and an uncanny food experiment.

To make this particular cheese, a "public" area of the body—the hand(s)—was chosen instead of more private ones so as to puncture some of the contradictions in human sociability. In its physical form the finished cheese pivoted around a fundamental contradiction: microbes are not-human when it comes to cleanliness (and therefore must be regularly and ritualistically sanitized out of our lives), but they are too human and too close for comfort in a body-cheese. In other iterations of body-foods, it seemed that the deliberate use of *human* microbes raised concerns, even outrage, in the case of feminist blogger @stavvers, who made bread leavened with yeast from her vagina (Another Angry Woman 2015). Food scientist Harold McGee (2014) reports about the microbiologist Stuart A. Koser, who made bread with bacteria isolated from an infected wound. While vaginas and flesh wounds transgress boundaries of personal health and hygiene, I chose to use hands as the swabbing site because the common nature of hands, handshakes, and the concomitant microbiota seemed accessible (and acceptable) to a public audience. By staging this cheese at a public tasting, the notion of eating—as both a literal and figurative form of embodiment—troubled the borders of self, other, microbe, and food.

At the crux of this provocation is the reality that handmade food is made all the time. Handmade food often connotes a heartfelt or artisanal gesture when the origins, intentions, or ingredients that "contain love" are at least known to contain a level of familiarity and trustworthiness (Fuchs, Schreier, and van Osselaer 2015); indeed the details imbued in handcrafted food marketing seem to contribute to the meaning-making processes of such gestures. Whereas fermented foods are also handmade, their microbial ingredients may not always register in the public perception of food production (e.g., in chocolate or coffee production), partly because this labor remains invisible. This cheese removed the "unknown" factor from the bacteria (i.e., the exact bacterial species remains unknown, but the presence of lactic acid bacteria is known due to the successful coagulation of milk proteins) and transferred it into the person whose hands were swabbed, thereby shifting the emphasis from "which bacteria?" to "whose hand?"[7] The hand (and its attached person) becomes the unknown variable and undermines the justification for controlling unknown microbes in kitchen/laboratory settings.

The invitation for the public to taste the cheese experiment exposed two other contradictions. Preparing the cheese would be like the work of chefs and countless other food artisans who also make "handmade" food but are rarely challenged. Furthermore, cheesemaking in the pre-Pasteurian era was done "by hand" without the hyperattention to sterilization and sanitation that characterizes the public health codes of today. Before commercial sanitizers were produced en masse, fermentation equipment was cleaned with boiling water, high-proof alcohol, or UV rays from the sun, all of which would kill off most, but not all, forms of microbial life. Not implausibly, the bacteria living on cheesemakers' hands would end up in the final product, especially since the temperatures of the milk, the cheesemakers' hands, and the lactic acid bacteria operate in a similar range (approximately 100°F). Historically—and in certain countries, still today—these unknown bacterial quirks contribute to the final tasting notes of a cheese, giving it a kind of complexity that reflects its distinct environment and maker. Similar to "open ferments" and "wild yeast" ferments, the microbial imprint of a particular place contributes to the flavor profile of the final ferment.[8]

The Making

What went into making the cheese, and how did it become credible, accepted, and actualized at a public event? It is worth noting that these experiments could not be conducted in the biology lab with which I am affiliated. Since its inception in 2016, the Speculative Life Lab at the Milieux Center for Arts, Culture, and Technology in Montreal has handled biological specimens ranging from human biological samples (*HeLa* cells, associated with cervical cancer), to pathogenic bacteria (*Serratia marcescens*, associated with fecal contamination) to benign bacteria that produce biofilms (*Acetobacter xylinum*, associated with kombucha). All of the lab's researchers, most of whom are graduate students and faculty, have completed the requisite training and institutional certification for operating in such a space. According to the golden rule of "no food in the lab," the milk was seen as a foodstuff and thus forbidden in the laboratory. These experiments could not be done in a commercial kitchen either, because the milk, now with the additional human microbes, was considered a biological specimen instead of a foodstuff. In these kitchen spaces food preparation with this milk-as-human-biological-sample was considered active contamination that made the cheese produced unfit for human consumption. What resulted was a double bind: working in one setting violates biosafety protocols, while working in the other setting violates food safety protocols.[9]

To reconcile these contradictions, the experiments were divided among three regulatory bodies. In order to make the cheese and serve it to a public audience, I applied for three institutional permissions: ethics approval (e.g., informed consent forms) to collect hand swabs, biosafety approval (e.g., protocols for a "handmade cheese" and biopermits) to make the cheese, and food waivers (e.g., allergen information) to properly serve the cheese. The piecemeal (and seemingly noncomprehensive) nature of this approach was most apparent in the partial guarantees that each regulatory body offered. Notably, the food waiver dealt only with the perishability of the milk. The biosafety protocol concerned itself only with identifying the microbial agents responsible for cheesemaking. The ethics review assured protection of the humans who provided samples, not the eaters per se. When combined, the limited scope of each regulatory body disappeared because each layer protected a different constituent in this provocation. Ethics concerned itself with how the sampling and swabbing were framed, so as not to harm or deceive persons offering a hand for swabbing. Food safety concerned itself with the eaters. Biosafety concerned itself with where the cheese was made. But no two domains addressed the same concerns of what constitutes a contaminant. Rather than overtly determining the definitions of a contaminant, this layered manner in which safety operated captured the complexity of food and the complexity of being human.

The "handmade cheese" did not harm the participants who offered their hands for swabbing (fulfilling ethics), and microbes known to be pathogenic were not used to make the cheese (fulfilling biosafety), which was properly stored and served (fulfilling food safety waivers). In its presentation to a scrutinizing public, the cheese exposed some of the contradictions we have in our beliefs about microbial life, and it opened up conversations around what those contradictions do.

The Tasting

This "handmade cheese" was made as a provocation to underscore permissible and impermissible ways of producing food and food knowledge. To increase the stakes, these cheeses were made using the hands of university staff members, and the cheeses were served to a public audience whose members could think through and taste the product(s). Four cheeses were produced for the public tasting event: three were made from the hand swabs of university staff members, with didactic panels indicating their occupation and their handedness. An additional cheese was made as a negative control; instead of hand bacteria, lemon juice was used to acidify the milk. The negative control was to provide attendees a low-stakes option and to offer a way of participating at different levels of engagement and interactivity. To minimize bias and avoid influencing people's decisions as they tasted the cheeses, the researcher remained on the opposite side of the room from where the cheeses sat. All cheese samples were stored on ice.

Responses varied, but the most common reaction was that of curiosity. Attendees noted the difference in taste across the samples, referencing the variability in flavor and texture across the three experimental samples of handmade cheese. One sample did not coagulate as much compared to the other samples, indicating very little microbial activity; it had a texture closer to that of yogurt. Another sample exhibited a sweeter taste, like mascarpone or almonds. By the end of the tasting event, the attendees had consumed the majority of the cheese samples.

This tasting event essentially asked attendees to participate in an experiment. In research-creation methodology the translated equivalence between the words *experiment* (English) and *éxperiénce* (French), calls for the experiential and the experimental to be addressed together. Experiments set up an encounter for us to experience complexity firsthand. The cultural studies scholar Ross Gibson (2010) explains the efficacy of provocations in a similar manner: "Why do these interactive, immersive and ever-emerging works matter? They matter because they give us a chance to sense directly how complexity works." In particular, the "handmade cheese" presents the complexity of food, of human bodies, and of our ideas about safety. The cheese made visible the invisible microbes that co-constitute our very corpora. The tasting acutely highlighted the microbial nature of fermented foods, but it also pointed to a larger ideological ferment about how humans always already entangle with microbial life.

Discussion

In the premise of the preceding case study, microbes are seen as a double threat: they threaten both the pure-human-body as well as the food that humans consume to replenish that sense of purity. The rhetoric used against microbes—*we must eliminate them*—does not hold when we consider ferments and how integral they are to our everyday metabolism and food culture. The next-best rhetoric—*we'll only let the good ones live*—remains problematic because valuating the worth of microbes (i.e., according to an exclusively human audience) reduces microbes to an essential function, tool, or resource to exploit and control. The essentialization of microbial life becomes apparent when their identity or purpose is unclear or when their "identity" or "purpose" is in flux based on context. Bacteria of the species *Lactobacillus delbrueckii*, for example, are hailed as friendly and health-affirming when found in probiotic yogurts; however, they are considered bad in beer production and pathogenic in instances of urinary tract infections. This cheese demonstrated that safety is not inherently fixed to a particular place, person, or set of practices but is instead emergent from their interactions.

Policing what counts as safe or contaminated relies on a troublesome set of purity politics, a kind of politicking that is built on top of asymmetrical notions of power and control. In her book *Against Purity* the social theorist Alexis Shotwell (2016) calls for a reimagining of collective ethics by dismantling the commonly held belief that we are individualistically and morally pure. She argues that purism sets up the ontological binary of "us versus them," and she cites the ideological allure of purity as "a common approach for anyone who attempts to meet and control a complex situation that is fundamentally outside our control" (8). The conflation of power and control applies to views of other species. Notions of human purity circle back on Pasteur, who "laid the groundwork for what [he] believed to be 'pure' social relations—relations that would not be derailed by microbial interruption, that could be predicted and thus rationally ordered" (Paxson 2008, 17). Purity is thus invoked in an attempt to reclaim (at least conceptually) the sense of control in instances of dealing with multiple, often invisible, forms of microbial life. These sentiments have been historically rooted, as noted by Anna Tsing (2012), who situates Pasteur's work within the political context of nineteenth-century European colonialism and the desire to take control over others' resources (e.g., rum) and production pathways (e.g., sugar plantations, the rum trade in the Caribbean). Control over colonies as a means to further the purist agenda hearkens back to fermentation revivalist Sandor Katz's observation on how bacteria are counted in numbers of colonies. In scientific nomenclature, microbes are counted in colony-forming units (CFUs). Perhaps the lexical consistency between microbial colonies and macrocolonialism points to similarly skewed understandings of power.

Undoing these assumptions about human power/control requires a shift in human primacy. Shifting the base of anthropocentrism means changing ethical stances and theorizing our interspecies relationality differently. In *The Origins of Sociable Life* Myra Hird (2009, 1) coins the term "microontologies," arguing for "a microbial ethics, or, if you will, an ethics that engages seriously with the microcosmos." Microbial ethics challenge the privileging of human stature through history and science. Whereas Darwinian evolution would view microbes and humans on a linear trajectory of increasing complexity (i.e., we evolved from microbe to plant to animal to human, with the human only distantly related to microbes in filial terms), alternate theories proposed by Lynn Margulis

(1998) argue that microbes are the basis for all life (i.e., both plants and animals should be viewed as hybridized forms of microbial life). Margulis contends that we are all microbial, adding that the term "we" functions as "a kind of baroque edifice" for our hybridized selves: "Our strong sense of difference from any other life form, our sense of species superiority, is a delusion of grandeur" (98). Hird's ethical reorientation and Margulis's evolutionary reorientation of human positionality coincide with broader questions about what it means to be human. More recently, the work of Ron Sender, Shai Fuchs, and Ron Milo (2016) updated the cell count of microbial cells in human bodies, indicating a one-to-one ratio and suggesting that we are just as microbial as we are human. Similarly, Thiago Hutter, Carine Gimbert, Frédéric Bouchard, and François-Joseph Lapointe (2015, 1) propose that humans are themselves multispecies since "human beings are so well integrated with their microbiomes that the individuality of human beings is better conceived as a symbiotic entity." Pointing to bee colonies that co-organize themselves as a superorganism, Hutter et al. reimagine humans as "emergent super-individuals" (2). Fermentation inserts itself into this problematic because its practices rely on cooperation across different scales of life.

Working in a fermentive space suggests that we—as both human and microbe—are not ontologically separate entities, interacting; rather, it proposes what the feminist philosopher and physicist Karen Barad (2007) names *intra*-action. The *intra* prefix connotes a nestedness within, suggesting that we are inextricably linked, co-composed, and share mutual stakes at hand. Compared to *inter*-action between two autonomous agents, intra-action erases the demarcation that defines selfhood and otherness. Fermentation, as an intra-active endeavor, does not consider human volition, microbial agency, and environmental factors as distinct things that interact to produce a ferment. Instead, intra-active fermentation locates agency in distributed terms. Thinking about fermentation as a process of Baradian intra-action challenges the dominant trend of rendering the human-microbe relationship in oversimplified terms. Hird applies the concept of intra-action to bacteria (that is, eschewing the "big like us" assumption about human actors) in order to examine our enmeshed and co-composed beings through digestion, immunity, and metabolism. Thus, intra-action extends the metabolism metaphor of interdependent exchanges to include the positionality of each (f)actor; it allows for a nested interpretation of humans, microbes, and spaces instead of keeping them isolated.

Fermentation productively engages with the unknown, the uncontrollable, and the emergent, yet an anthropocentric focus limits knowledge production in the kitchen/laboratory contexts. Rather than privileging one set of food truths, fermentation encourages multiplicitous results to exist simultaneously because so much of the fermenting process is predicated on an ever-changing ambient. Temperature, humidity, salinity, acidity, and porosity construct the spaces for certain species to thrive and others to desist. Thought of this way, fermentation is a process of *working with* spatial affordances. It is necessarily connective, showing the inseparability of humans, microbes, foods, and the spaces we collectively enact.

Conclusion

This chapter has sought to move beyond the figurative and symbolic role of fermentation in society, to literally shift our ideas about how we engage with microbial life in kitchen spaces, laboratory spaces, inside our bodies, and outside them. A case study was presented in which milk became unknown through its fermentive transformation into a "handmade cheese." This transformation from known to unknown, from foodstuff to uncanny experiment, confronted participants with the notion that there are "more of us" at work inside kitchen/laboratory spaces. The intervention exemplified the combined politics of operating in kitchen/laboratory spaces, thus serving as a case study for how to improve our understanding of the shifting power relations in fermentation.

Working with fermentation in kitchen/laboratory spaces shifts the ideological primacy of human positioning, recasting the human instigator as neither the top nor the center of a food space. Fermentation suggests an intra-active relationship with microbes, one in which humans and more-than-human actors collectively work with spatial affordances to bring about fermentive change. As a heuristic for material and semiotic change, fermentation challenges who or what is granted agency in these spaces, because these considerations influence what is ultimately produced.

Notes

The author would like to acknowledge Chelsea Leiper and Colleen Myles for their feedback on initial drafts, as well as Shadi Maleki on subsequent formatting. This chapter also benefited greatly from the comments of anonymous reviewers. Thank you.

1. Consider how fermentation transforms grapes into wine. This change is biochemical (fructose to ethanol, for example), and the resulting ferment carries new meanings that were not associated with the original grape (e.g., hospitality, religious symbolism).

2. This chapter is meant to be illustrative, not exhaustive, in its mobilization of archetypal kitchens and laboratories. To be sure, not all kitchens and not all laboratories behave as described, and hybrid spaces (such as the Nordic Food Lab, Basque Culinary Center, or various restaurants' fermentation departments) defy these descriptions. Nevertheless, this chapter makes a comparison of spaces that produce (food) knowledge insofar as it brings to the fore questions of human positionality in the context of food-making and food research.

3. This begs the question of whether theory or practice preceded. Did science's obsessions with consistency become the warrant for sterilization? Or did sterilization enable consistent findings to occur, thus validating its use?

4. The anthropocentric perspective here should be made apparent, for we (as *anthropos*) often take such perspectives for granted. While "foodborne illnesses" inflict harm to the human host, the microbe is thriving quite well!

5. See the book by Lynn Margulis (1998) on the origins of mitochondrial DNA.

6. See the installation by Agapakis and Tolaas (2013), a collaboration exhibited at the Dublin Science Gallery. According to the gallery's website, it is "a series of 'microbial sketches,' portraits reflecting an individual's microbial landscape in a unique cheese."

7. The relationship between visible/invisible labor and food acceptability would be interesting to scrutinize at larger scales of life. For example, *Aspergillus* spp. fungi are often used in high-fructose syrup production (e.g., high-fructose corn syrup) but remain hidden from public perception. Large-scale concentrated animal feeding operations (CAFOs) also capitalize on invisible labor (e.g., artificial insemination,

media censorship), but their unacceptability rests on the moral compass of individuals (versus systemic moral rejection).

8. Unpasteurized milk is regulated precisely because of its "unknown" microbial profile. Mother Noella Marcellino, a trained microbiologist and Benedictine nun, studied the traditional French methods for making fresh cheeses (like brie and Camembert) from raw, unpasteurized milk and proved that these methods allowed naturally occurring microbes in raw-milk cheeses to fight off potentially harmful ones. The resulting panoply of raw-milk cheeses available in France proves that the absence of pasteurization does not necessarily pose a threat. Rather, the biodiversity of microbes in unpasteurized cheese turns into richness and depth of flavors. See also Paxson's (2008) work on raw-milk cheese in Vermont.

9. I ultimately made the cheese in my own home kitchen after filing for a biosafety permit. This poses interesting questions for home spaces as terrains for knowledge production.

References

Agapakis, Christina, and Sissel Tolaas, artists. 2013. *Selfmade*. Installation at Dublin Science Gallery, Trinity College, Dublin, Ireland.

Another Angry Woman. 2015. "Baking and Eating #cuntsourdough" (blogpost), November 5, 2015. https://anotherangrywoman.com/2015/11/25/baking-and-eating-cuntsourdough/.

Barad, Karen. 2007. *Meeting the Universe Halfway: Quantum Physics and the Entanglement of Matter and Meaning*. Durham: Duke University Press.

Chapman, Owen, and Kim Sawchuk. 2012. "Research-Creation: Intervention, Analysis and 'Family Resemblances.'" *Canadian Journal of Communication* 37 (12): 5–26.

Fuchs, Christopher, Martin Schreier, and Stijn M. J. van Osselaer. 2015. "The Handmade Effect: What's Love Got to Do with It?" *Journal of Marketing* 79 (2): 98–110.

Gibson, Ross. 2010. "The Known World." Special issue website series, *Text* (8).

Hird, Myra. 2009. *The Origins of Sociable Life: Evolution after Science Studies*. Basingstoke: Palgrave Macmillan.

Hutter, Thiago, Carine Gimbert, Frédéric Bouchard, and François-Joseph Lapointe. 2015. "Being Human Is a Gut Feeling." *Microbiome* 3:9. https://doi.org/10.1186/s40168-015-0076-7.

Latour, Bruno. 1988. *The Pasteurization of France*. Translated by John Law. Cambridge MA: Harvard University Press.

Loveless, Natalie. 2012. "Practice in the Flesh of Theory." *Canadian Journal of Communication Studies* 37 (1): 93–108.

———. 2015. "Introduction." *RACAR: Revue d'Art Canadienne/Canadian Art Review* 40 (1): 41–42.

Margulis, Lynn. 1998. *Symbiotic Planet: A New Look at Evolution*. New York: Basic Books.

McGee, Harold. 2014. "The Disquieting Delights of Salt-Rising Bread." *Popular Science*, May 20, 2014. https://www.popsci.com/article/science/clostridium-it-can-kill-you-or-it-can-make-you-bread.

Paxson, Heather. 2008. "Post-Pasteurian Cultures: The Microbiopolitics of Raw-Milk Cheese in the United States." *Cultural Anthropology* 23 (1): 15–47.

Sender, Ron, Shai Fuchs, and Ron Milo. 2017. "Revised Estimates for the Number of Human and Bacteria Cells in the Body." *PLoS Biology* 14 (8): e1002533.

Shotwell, Alexis. 2016. *Against Purity: Living Ethically in Compromised Times*. Minneapolis: University of Minnesota Press.

SSHRC (Social Science and Humanities Research Council). 2016. "Definitions of Terms: research-creation." http://www.sshrc-crsh.gc.ca/funding-financement/programs-programmes/definitions-eng.aspx#a22.

Tsing, Anna L. 2012. "Unruly Edges: Mushrooms as Companion Species." *Environmental Humanities* 1 (November): 141–54.

Zymurgeography? 11

Biotechnological Ferments and the
Risks of Fermentation Fetishism

Andy Murray

Fermentation Matters

In the biotechnological age the manipulation of living matter continually transforms humans' relationships to the more-than-human world and opens up new possibilities for production. While the science of genetic engineering that typically exemplifies biotechnology in the twenty-first century is fairly new, fermentation, which humans have fostered for millennia, could be considered the first "biotechnology" (Bud 1993). Beginning with Louis Pasteur and extending into the twentieth century, fermentation became biologized, and what we now recognize as biotechology evolved in tandem with fermentation and fermentative industry. Now, when technologies of genetic engineering are more precise and more capable than ever, innovations in the biotechnology of fermentation continue to transform fermentative production. Unlike some of the earlier biologically informed adjustments to fermentation, these transformations affect more than efficiency and reliability. By altering the genetic code, protein expression, and metabolic pathways of microorganisms, they more fundamentally alter fermentation itself, expanding both the biological definition of the word and the role that it can play in transforming landscapes. These ferments produce more than alcohol and the other foodstuffs with which the word "fermentation" is still usually associated in common parlance; they produce fuels (and not just ethanol), pharmaceuticals, and sophisticated food substitutes.

Echoing some of the sentiments from when fermentation first became an object of biotechnological fascination in the mid-twentieth century, those in the new fields of synthetic biology and metabolic engineering have touted fermentation as capable of addressing some of the world's most pressing issues. They position fermentation as capable of not just breathing new life into agricultural landscapes and lifeways but also dealing with energy and medicine shortages and the environmental and animal welfare tolls of industrial dairy.

The contributors to this volume experiment with using fermentation as a metaphor and a way of theorizing landscape change. They are not alone in identifying fermentation's conceptual appeal. Other social scientists have recently taken to thinking of fermentation as a useful metaphor (Bobrow-Strain 2012; DuPuis 2015), particularly one that suggests a theoretical or political orientation that rejects untenable notions of purity and instead favors imperfection and constant change (Shotwell 2016). Although fermentation can be a productive metaphor to think with, it can also be unwieldy. And like the microbial metabolic process itself, fermentation's meaning is effectively inflected by those in the natural and engineering sciences. The shifting nature and meaning of fermentation contribute to discrepancies of understanding. While flows of and through microbes and their metabolisms are manipulated more quickly and more profoundly than ever before, other forms of fermentation—its traditional or "craft" forms—soar in popularity and visibility, and fermentation gains currency as a theoretical tool in the social sciences. These concurrent trends result in disparities between the ways in which natural scientists and engineers, typical consuming publics, and social scientists—who are perhaps just now beginning to take a serious interest in microbial matters—understand what fermentation is and does. In the politically fraught worlds of biotechnology and production, such disparities run the risk of setting the scene for future conflict.

This collection represents both a call to pay attention to the ways in which fermentative industry transforms landscapes and an exploration of whether fermentation provides a unique lens for theorizing landscape change, whether that change is specifically related to fermentative industry or not (see chapter 13, this volume). Both noble aims face risks related to a failure to engage more closely with fermentation's complex history and diverse—and diversifying—forms. Drawing on insights from science studies, analysis of synthetic biology, and metabolic engineering discourse, as well as findings from ethnographic fieldwork among "biohacking" community-based synthetic biologists, I argue that fermentation's essential complexity and contestedness pose significant challenges for attempts to understand it in any general sense or to apply it as a metaphor. Deriving theory from only certain of fermentation's forms can result in what I call "fermentation fetishism," which itself poses further risks worth considering. First, it may merely draw convenient parallels to fermentation by defining its scope based merely on forms that make a nice conceptual fit with existing ways of theorizing landscape change. In other words, it may simply be that certain of fermentation's forms are compatible with how we already think about landscape change, in which case its use as a metaphor is of little benefit. Second, and I think more important, is that it may contribute to an already prevalent narrow focus on familiar, traditional, or trendy forms of fermentation at the expense of developing a broader understanding of how fermentation's newest technologized forms also stand poised to transform landscapes in profound ways. Put succinctly, new meanings and applications of fermentation can come at the expense of a better understanding of fermentation itself.

Like Marx's commodity fetishism, this fetishism involves both the impression of equivalence and a collective forgetting or unawareness of the underlying grounds for this equivalence (see Marx 2004). For Marx, this superficial equivalence comes in the form of the exchange value of different commodities and at the expense of a forgetting of the essential incommensurability of these commodities' use-values and the origins of this value in labor. In this case it involves comparisons between various forms of fermentation and may come at the expense of attention to important process details: the bases for technical comparison, the ways in which bioengineering is pushing the limits of such comparisons, the human cultures of fermentation, and biotechnology's history and future likelihood of eluding control and inciting controversy. This fetishism already appears in fermentation discourse, and its presence showcases some of the dangers it poses for building better collective understandings of technoscience and production. If using fermentation as theory or metaphor is to avoid these dangers, it will necessitate remaining mindful of how bioengineered fermentation will likely continue to grow its capabilities and influence as the nascent field of synthetic biology grows and matures and the biological sciences tighten their embrace of engineering epistemologies (Roosth 2017).

To complicate a narrow or superficial conceptualization of fermentation, I begin by providing a brief history of fermentation as a biological and biotechnological object. Then I discuss the more recent rise of bioengineering, synthetic biology, and metabolic engineering and how they have transformed fermentation still further and produced a streak of fermentation fetishism.[1] I draw on ethnographic observations and interviews from grassroots DIY or community biology laboratories (or "biohackerspaces"), as well as select fermentation and metabolic engineering textbooks and academic articles.[2] My goals are to understand how fermentation is manipulated and deployed and to understand the implications and effects of these developments for understanding fermentation's potential, both for landscape transformation and as a useful metaphor for transformation itself—and to explore whether these two aims are at odds.

From "Zymotechnology" to Synthetic Biology: A Brief History of Microbial Technoscience

The English word *fermentation* comes from the Latin *fervere*, which means "to boil." As this etymology indicates, the word first referred to the bubbly activity of alcohol and pickle production, the microbial source of which was unknown at the time. Brewers would recycle the slurry from fermentation tanks, recognizing its importance to the fermentation process and referring to it by such names as *yeste* and *godisgood* (White and Zainasheff 2010, xvi).[3] A debate developed about the underlying causes of fermentation, particularly as to whether it was a spontaneous process or one that was purely chemical or biotic.[4] Pasteur sought to settle the matter by devising controlled experiments using a sealed apparatus filled with a fermentable broth and sterilized with heat, ultimately demonstrating to the scientific community's satisfaction that fermentation was a biotic process carried out by single-celled organisms. Pasteur's studies culminated in a book, *Studies on Fermentation: The Diseases of Beer, Their Causes, and the Means of Preventing Them*, and his efforts definitively placed fermentation in the domain of the life sciences.[5]

The histories of modern fermentation and biotechnology are entangled well beyond Pasteur's path-breaking work. Fermentation deeply informed the original practical and conceptual development of biotechnology. Robert Bud (1993) describes the historical transition from fermentation technology, or "zymotechnology," to "biotechnology."[6] He explains how biotechnology, which eventually developed into a distinct field of technoscience combining conceptions of life developed within the still-young discipline of biology with engineering's emphasis on intervention and practical application, developed from a more limited set of practices concerned specifically with fermentation. Following Pasteur's breakthrough, fermentation grew into a large-scale industry as brewers hired biochemists to improve their products. They developed techniques for culturing pure strains of microorganisms, and modernized industrial fermentation and biochemistry evolved largely in tandem.

In the early twentieth century the word *biotechnology* first entered the English lexicon, and the fusion that it denoted informed both the theory and practice of working scientists (Bud 1993). Since that time biology—particularly at the cellular and molecular levels—has in many ways become an engineering discipline. At first connected with selective breeding and genetic improvement programs and later with genomics and more precise techniques of genetic modification (GM), biological engineering had many noteworthy achievements and developments in the twentieth century: artificial parthenogenesis (Pauly 1987); mammalian cloning and other advanced reproductive technologies (Thompson 2005; Franklin 2007); the polymerase chain reaction (PCR) (Rabinow 1996); full genome sequencing (Hayles 1999; Keller 2002); advanced techniques and technologies of cell culture (Landecker 2007); patentable genetically engineered organisms, including not only microbes but also plants and animals; and more precise gene-editing techniques, including the much-publicized CRISPR-Cas9, a remarkably precise and inexpensive genome-editing technology.[7] Through these and other developments, bioengineers have increased their ability to manipulate life at the cellular level and thereby rendered these organisms increasingly, albeit selectively, plastic (Landecker 2007; Murray 2018).

While the range of current bioengineering projects is broad, fermentation remains a major part of biotechnology's present. The emergence of the field of so-called "synthetic biology" has reinforced the presence of a tinkerer's epistemology—the conviction that the best way to know life is to manipulate it—in the biosciences (Calvert and Martin 2009; Calvert 2013; Roosth 2017). Spurred in part by the migration of a large number of PhD holders from engineering disciplines—including chemical, mechanical, and computer engineering—to the biosciences, synthetic biology focuses on building artificial biological organisms and systems. One major area of work among the several in synthetic biology—including, for example, building biological computers and creating de novo synthetic life forms—is the improvement of metabolic engineering (Stephanopoulos, Aristidou, and Nielsen 1998), an area of specialty that inherits the legacies of zymotechnology and chemical engineering and focuses on the manipulation of microbial metabolisms to produce high-value substances (Roosth 2017). Through efforts to use microbes as tiny protein factories and their metabolic pathways as production lines, synthetic biology and metabolic engineering are actively remaking and redefining fermentation.

According to what is probably still its most common biological definition, fermentation refers specifically to a metabolic pathway that produces ATP (adenosine triphosphate, or cells' primary energy source) in the absence of oxygen, an anaerobic alternative to oxygen-dependent cellular respiration. While ATP may be the cell's incentive for fermenting, the process also creates other (by)products: lactic acid (think sauerkraut or the "burn" of your muscles during intense exercise) or ethanol and carbon dioxide (think beer or champagne). As a result of the rise of biotechnology and the epistemic authority of its practitioners in defining biological concepts, this definition is now changing to include other processes, including modified protein expression and engineered metabolic pathways, both of which result in novel ways to produce desirable substances. According to the common definition of fermentation, these substances are beyond its productive capabilities. They include, for example, complex proteins like biopharmaceuticals and fragrances.

The present-day practice of modifying bacteria and yeasts to produce new desirable substances stretches back at least to the discovery of penicillin. This discovery opened up the possibility of producing large quantities of new substances using microbes as an input-output system. Before long, several different antibiotics were being produced in large quantities. Drawing on these innovations, Herbert Boyer, cofounder of then-young Genentech, in collaboration with California's City of Hope National Medical Center, developed a way to make an incredibly prized substance—human insulin—using similar methods (Genentech 1978).[8] Stanley Cohen and Herbert Boyer's (1980) patent for the production of insulin by genetic engineering has this to say about the technology of turning microbial metabolisms into productive forces: "Various unicellular microorganisms can be transformed, such as bacteria, fungii [sic] and algae. That is, those unicellular organisms which are capable of being grown in cultures of fermentation."

As genetic engineering became more established, academics and entrepreneurs began to use retooled microbial metabolisms to produce still more high-value substances. These developments were fueled both by the success of Genentech and their insulin and by major technical developments like the polymerase chain reaction, which allows the reproduction of large quantities of DNA (Rabinow 1996), and later CRISPR-Cas9, which allows cheaper precise gene editing than ever before. Ethanol of course was perhaps the earliest intentional product of human-directed microbial metabolism, but synthetic biologists applied bioengineering techniques to produce a range of other "biofuels," including a gasoline analog designed to replace car engines' combustible of choice without the corresponding need to change existing engine designs and fuel delivery infrastructures (so-called drop-in "biogasoline") (Foo et al. 2014). Synthetic biology has begun to gravitate toward the production of foods and fragrances (Hayden 2014), such as food additives and flavorings like vanillin or a sophisticated milk substitute suitable for vegans (Perfect Day 2016b). In significant biopharmaceutical breakthroughs, synthetic biologists have managed to use microbes to produce the antimalarial drug precursor artemisinic acid (Ro et al. 2006) and hydrocodone, an opioid (Galanie et al. 2015).

Fermentation textbooks that introduce the subject to students make clear that as bioengineering expands the range of microbial metabolic products, scientists and engineers have started referring to many different metabolic processes—not merely the anaerobic pathway of the classic biological definition—as "fermentation." Peter Stanbury, Allan Whitaker, and Stephen Hall (2014) place all fermentation practices on a continuum, from alcohol and vinegar production pre-1900 through the post-1979 genetic engineering and the production of "foreign compounds" using microbial cells. As the authors explain, "The production of alcohol by the action of yeast on malt or fruit extracts has been carried out on a large scale for very many years and was the first 'industrial' process for the production of a microbial metabolite. Thus, industrial microbiologists have extended the term fermentation to describe any process for the production of product by the mass culture of a microorganism" (Stanbury, Whitaker, and Hall 2014, 1).

Another textbook opens by stating that "fermentation has been known and practiced by humankind since prehistoric times, long before the underlying scientific principles were understood" (El-Mansi 2012,

1), and proceeds to describe how microbiology first shed light on and then eventually modified these ancient practices through bioengineering. By remaking microbial metabolisms, bioengineering is redefining fermentation itself. Fermentation is not merely a process of flux; it is also a process *in* flux. These changes have implications for anyone wishing to understand or use fermentation, including those who want to understand its transformative effects and metaphorical utility.

Something New Is Brewing: Engineering and Crafting Equivalence

The flux of fermentation, and the ways in which the word now unites more diverse efforts than ever under the same nominal umbrella, is significant because it produces a conceptual ambiguity that may impede communication and understanding among natural scientists and engineers, social scientists, and consuming publics. One example of this ambiguity in action is how bioengineering advocates skirt potentially controversial elements of their production process through superficial comparisons to other, more traditional or familiar forms of fermentation. Frequent comparisons among types of fermentation that create accessible and generally nonthreatening analogies for bioengineering are based on select technical elements, while also taking advantage of positive associations that are nontechnical in nature. This reinforces a superficial equivalence between different forms of fermentation and largely ignores or downplays other elements, like genetic engineering and the fact that many of the activities of synthetic biology and metabolic engineering only recently became classifiable as "fermentation" at all. Occasionally this practice veers into a more specific fermentative comparison, one example being brewing—or, even more specifically, "craft" brewing. This is not especially surprising given the intertwined histories of beer and microbiology, as well as the fact that brewing is a long-established technology and a popular and relatable touchstone for many consumers otherwise unfamiliar with fermentation. Using it as such a touchstone, however, also leverages support for what in the modern age are only loosely related forms of production, implicitly suggesting shared practices and values when in reality few may exist. It also fails to do justice to the diversity in brewing practices and values, which are already a site of both practical ambiguity and political struggle. The result is a gap in understanding and an analogy that risks yielding less clarity rather than more.

In the scientific literature many metabolic engineers describe their work as "fermentation," using the term's newer biological definition. Jay Keasling, a prominent synthetic biologist and metabolic engineer, is known for his production of biofuels and engineering baker's/brewer's yeast (*Saccharomyces cerevisiae*) to generate the antimalarial drug precursor artemisinic acid (Ro et al. 2006; Foo et al. 2014). His publications assert that "combining . . . natural fatty acid synthetic ability with new biochemical reactions realized through synthetic biology has provided a means to divert fatty acid metabolism directly towards fuel and chemical products of interest" and thereby "produce these products directly from abundant and cost-effective renewable resources by fermentation" (Steen et al. 2010, 559) and describe his efforts "to inexpensively produce the antimalarial drug artemisinin through a new fermentation process" (Hale et al. 2007, 198). Fellow biofuel producers Yanfeng Liu et al. (2015, 1109) argue that "significant socioeconomic benefits of microbial fermentation, such as environmentally-friendly processes and sustainability, have drawn the attention of researchers seeking to develop fermentation methods for chemical and fuel production." Christina Smolke is known for the production of opioids using metabolic engineering and also edited *The Metabolic Pathway Engineering Handbook*. Her publications describe her work as follows: "fermentation with engineered yeast is a scalable platform for production of complex plant alkaloids" (Galanie et al. 2015, 144) and "we developed baker's yeast . . . as a microbial host for the transformation of opiates . . . performing high-density fermentation" (Thodey et al. 2014, 837).

While those reading these papers are most likely knowledgeable about the practical differences between metabolic engineering and other ferments, in more public-facing communications metabolic engineers' technical references to fermentation can veer into specific comparisons to

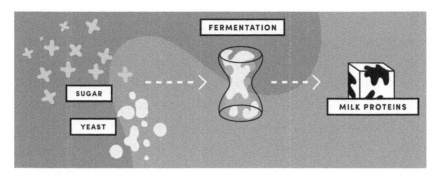

FIG. 38. This graphic depicting Perfect Day's production process shows how the term "fermentation" is used as descriptive shorthand for a complex production process that includes genetic engineering and departs from the conventional biological definition of the fermentation process. Courtesy Perfect Day.

traditions of food production and brewing. Keasling, in a CNN (2013) interview with Sanjay Gupta, said, "We use it [yeast/fermentation] to make bread, we use it to make beer, we use it to make wine. And it's been used for centuries."[9] San Francisco startup Perfect Day (formerly Muufri) provides a clear example of this type of comparison, and one worth exploring in depth, for the ways in which it showcases fermentation fetishism by skirting discussion of technical details and using fermentation as a shorthand signifier for its production process.

Perfect Day (2016b) explicitly compares its work making a vegan milk substitute to "craft" brewing: "Instead of having cows do all the work, we've developed a process similar to craft brewing. Using yeast and age-old fermentation techniques, we make the very same dairy proteins that cows make." The company reasserts but does little to expand the comparison in their FAQ section: "[Q:] Is your process kind of like making beer? [A:] Part of our process relies on old-world fermentation techniques—similar to making beer. We often describe it as dairy meets craft brewing" (Perfect Day 2016a).[10] Their mock-ups of bottles pictured on their website even say that their milk is "brewed with love in San Francisco" (Perfect Day 2016b).

These comparisons between recent bioengineering work and "Old

World" or centuries-old forms of fermentation like brewing reinforce conceptual connections enabled by the fluidity of fermentation but wade even further into ambiguous and tenuous equivalence. This targeted comparison positions bioengineering as akin to the traditional forms of food and alcohol production but comes at the expense of accurately or thoroughly characterizing either bioengineered or alcoholic ferments. Notably, Perfect Day's focus on this particular comparison also downplays other elements of their process, particularly the genetic engineering aspects and accompanying controversial associations with genetically modified organisms (GMOs). Considering the controversy that has often surrounded these technologies, reluctance on the part of consumers to accept GM methods for a broader array of foods and fragrances could prove a substantial obstacle to success (see Shiva 1997; Schurman and Kelso 2003; Jasanoff 2005; and Parthasarathy 2017). In the United States and perhaps even more so in Europe, consumers have proven wary of claims that GM technologies are desirable or their products safe for human consumption, despite broad scientific consensus that they are (National Academies of Sciences, Engineering, and Medicine 2016). Many of my research participants who are themselves engaged in genetic modification interpret this wariness as indicative of a lack of trust or knowledge and a conflation of the different types of, uses for, and threats posed by GMOs (interview, October 17, 2018; interview, October 15, 2018; interview, August 8, 2018).

In acknowledgment of the ways in which consumer concern seems to have focused increasingly on GMO consumption, Perfect Day provides assurance that no genetically modified organisms will be present in finished products. This briefly addresses concerns over the effects of GMO consumption while ignoring other facets of their creation and use. Further, while one can find this information on the company's website, it is relegated to their FAQ section, as part of a negative response to the question, "Does your milk contain GMOs?" (Perfect Day 2016a). Their homepage meanwhile describes this process as simply "cultivat[ing] . . . yeast to produce dairy proteins" and as using "age-old fermentation techniques" (Perfect Day 2016b). In this way, leveraging comparisons to "age-old" fermentation presents the technology in a way that avoids any discussion of some of the process's more controversial elements. These elements include other drivers of public backlash against GMOs—like the aggressive use of intellectual property protections or large capital interests' growing control over food production—as well as the broader landscape changes, both social and physical, that would result from reorganizing existing "ecologies of production" (Paxson 2012, chap. 2) around bioengineered microbial metabolisms. Overall, the comparison to traditions of fermentation and cultivation limits the scope of activities under consideration.

Perfect Day's comparison implies commonality with a seemingly specific and currently popular form of fermentation: craft brewing. But what makes their process like "craft" beer production? A couple of different sources provide clues as to how difficult the idea of "craft" is to pin down. The Brewers Association (BA), which certifies brewing operations with its "Independent Craft" certification and seal, defines craft beer producers based on the criteria of production volume, independent ownership, and the apparently all-inclusive focus on "traditional or innovative brewing ingredients" (Brewers Association n.d.). Discussing the concept beyond the beer industry and drawing from a variety of sources, Colin Campbell (2005) notes that "craft" production often carries associations of small scale, design oversight, and authenticity.[11]

Vague as these criteria are, almost any new metabolic engineering endeavor can claim similarity to craft brewing based on them. Production volumes for experimental ventures are going to be small, even if their ultimate aspirations are large. Startup companies like Perfect Day and craft brewing operations alike tend to be little. They are staffed by a small number of persons and generally overseen by just a few. The ownership criterion for craft breweries is to indicate that they are not majority-owned by the small collection of massive "macro" brewing corporations. Building a different kind of "brewing" operation using a venture-capital startup model also matches the criteria. The last BA criterion is both vague enough to apply almost regardless of brewing operation and representative of precisely the line that Perfect Day and others who compare bioengineered ferments to "age-old" forms of fermentation try to walk. Fermentation is a traditional process, this is merely the latest innovation, and it's an innovation to re-create a familiar product. Relatedly, authenticity is a difficult concept to judge in any case, especially in any industry that openly praises innovation. While BA does provide a few other, lower-tier "concepts related to craft beer and craft brewers," these are also vague and generally reinforce the three major criteria.

Overall, the fact that the definitions of "craft" and "craft beer" are, like that of fermentation, fluid and a bit unclear helps Perfect Day's case. Even if the fit is loose, who's to say that what they are doing isn't indeed "dairy meets craft brewing"? Perfect Day's approach takes advantage of conceptual fluidity in more ways than one and shows how taking advantage of such muddy concepts creates the impression of similarity and circumvents potentially ethically or politically fraught topics, which may also include a deeper elaboration of the technical processes involved and the social networks—commodity chains, production skills, and regulations, for example—required to sustain them.

Differences That Make a Difference:
Fermentative Ethos and Arts of Distinction

Comparisons between forms of fermentation can be less confounding when they engage the technical and practical distinctions between different forms of fermentation. Rather than focusing on superficial likenesses, these efforts more openly take on questions of technics (technical details), scale, and economics and thereby provide foundations for less fetishistic and more substantive discourse about fermentation's different forms. Drew Endy, Stephanie Galanie, and Christina Smolke (2015, 2) provide one such example, pointing out that "there are differences between industrial bioreactors and 'home-brew' fermentation." They conducted an experiment that illustrates how comparisons between different forms of fermentation break down in practice, and they did so by attempting to use a yeast engineered to produce an opiate, thebaine, under homebrew fermentation conditions. Their results suggest that the practical differences between forms of fermentation really matter—in other words, they are differences that make a difference (Bateson 1972).[12] According to Endy, Galanie, and Smolke (2015, 2), "We used yeast that make an English ale as a positive fermentation control. We observed no production of thebaine and miniscule [sic] amounts of reticuline, an upstream biosynthetic intermediate, in home-brew fermentations; the positive control was palatable. We suggest that additional technical challenges, some of which are unknown and likely unrelated to optimized production in large-volume bioreactors, would need to be addressed for engineered yeast to ever realize home-brew biosynthesis of medicinal opiates at meaningful yields."

The authors engaged in this experiment primarily to assuage concerns that some of their colleagues had expressed about the feasibility of home-brewing opioids in the midst of a widespread opioid addiction crisis. In other words, their colleagues expressed concern stemming from the prospect of homebrewed opioids, and only in light of this concern was it revealed that there are crucial technical elements of different fermentation processes that are not only substantively different but even "likely unrelated." The potential for controversy, then, can provoke a more considered analysis of similarities and differences, if it becomes pressing enough (or perhaps if it becomes enough of a threat to the success or profitability of a given project). This suggests the need to insist on the ethical and moral dimensions of fermentation technologies to highlight the need for more substantive discourse. In addition to social scientists and bioethicists, one group that has worked hard to insist on these dimensions is "DIY-biologists," also known as "biohackers."[13] Their work helps show that using fermentation as an analogy does not necessarily obfuscate either technical or ethical matters, even though it risks doing so.

Due to decreased costs and the perceived potential of synthetic biology, a number of community laboratories that engage in DIY-bio, or biohacking, have emerged as part of the larger synthetic biology movement. At these labs the historical entanglement of biotechnology and fermentation remains on display. Community biology laboratories, including the ones I have worked with, regularly hold events and classes related to "traditional" products of fermentation: kombucha, sauerkraut, or mead. While one community lab's fermentation instructor views the synthetic biology work in the lab as drastically different, much more "precise," and finicky (and even feels this way about certain types of beer production, which he jokingly calls "monk-y brews"), the fermentation classes do draw people into the laboratory and help indirectly introduce them to lab work and bioengineering (interview, April 18, 2018; field notes, September 27, 2017). Participant observation and interviews from community bio labs reveal how and why some of their work constitutes an ethical critique of other forms of fermentation and requires them to delve into practical distinctions. While not true of all biohacking projects, some DIY-biologists pursue genetic engineering not for profit or efficiency's sake but because they understand it as a way of "doing politics" and addressing social problems—including problems with bioengineering as currently practiced (interview, December 10, 2017; interview, January 14, 2018). While biohacking revolves around bioengineering and lab work, community biologists often adopt a specific orientation to, and broader view of, this work.

Like some of their counterparts among the community of synthetic biologists, some biohackers may also make comparisons to brewing as a way to communicate their work and mission, as well as to contextualize it in relation to other microbial production practices. According to one member of a community lab working on producing an open-source generic pharmaceutical using bioengineering,

> With brewing beer there's a very direct set of correspondences with what we're doing. We're growing yeast, and feeding it in sort of a liquid medium, and we're monitoring the conditions of that, and we're trying to keep it clean and free of contamination, and then whatever comes out of

that, we're going to have to take something out of it, and purify it and bottle it up, and that is pretty much what you do at a brewery. Purification in particular is probably a little more complex than brewing beer. . . . It's very similar. And if you tell people, "We're going to create, like, [a pharmaceutical] microbrewery" or something, then it's a lot easier to contextualize that, because it provides you with this whole other set of associations that people are much more familiar with. Where you can say, there used to just be Coors and Busch, and we want there to be a couple in every city now. And we're working with yeast, we're fermenting things, it's a lot—not exactly like brewing beer—but it's a lot like it. So you can imagine instead of a tank as big as a building, you just have a tank that is like what you would do in your garage to brew beer. (interview, December 10, 2017)

While this characterization does focus on loose similarities between types of fermentation, it also acknowledges some of the different practices of fermentation even within the realm of beer brewing. Biohackers compare their work to some kinds of fermentation working against others, rather than using fermentation itself as shorthand for a nebulous productive ethos, as Perfect Day or Jay Keasling seem to do. While they pursue small-scale, disruptive, and decentralized production using fermentation, they largely avoid fetishizing fermentation as inherently any of these. Put differently, while these community labs do seek technological fixes for social issues, they are also driven to acknowledge the deeply socially embedded nature of these technologies by virtue of their orientation against what they perceive as the abuses of technically similar endeavors. They engage in genetic engineering, but they also foreground concerns with the ways that genetic engineering efforts have leveraged patents and other intellectual property protections in the past (interview, December 10, 2017; interview, January 14, 2018). Furthermore, rather than seizing on a gap of understanding between bioengineers and "lay" publics, they approach the rapid advancement of bioengineering and synthetic biology as an opportunity for reimagining the connections between biology and publics more broadly. One lab member summarized this overarching mission: "the role of the community lab is to make [science] palatable for a bigger audience, where it becomes something that occurs more frequently in daily parlance, [and] people develop a fluency" (interview, October 10, 2018). Given their modest successes with production so far, these laboratories are perhaps less involved in producing substances through fermentation than in producing critiques of existing institutions and supply chains. They use fermentation as an analogy to highlight issues of scale, centralization, and ownership.

These community laboratories could provide a good point of entry for broader engagement, including of social scientists, into the area of biotechnologized fermentation, its pursuit of value, and its potential for controversy. Other efforts by social scientists to participate in synthetic-biology-in-the-making—including working closely with Jay Keasling—have proven frustrating, largely due to the inflexibility of existing institutions (Rabinow and Bennett 2012). In contrast, some DIY-bio labs' open-door policies and willingness to accept walk-in community members and biological nonexperts and to entertain their questions is promising. However, while they do this in good faith, it does not necessarily translate to meaningful participation or the opportunity to reshape existing projects, which in my experience are often still primarily technologically driven and revolve around a relatively small core group of people.

The self-selected nature of laboratory participation, interests in expediency, and the need to demonstrate progress can also result in skirting controversial issues. For example, one member of a lab that runs a project to bioengineer a vegan food substitute admits that they have been fortunate to deal with few complaints about GMOs: "a lot of people who might be concerned about genetically modified organisms in general also may be vegans or are sympathetic to that sort of thing, so that's also really fortunate" (interview, December 10, 2017). They generally do not shy away from the genetic engineering elements of their work, but they do at times view it as a misplaced concern and a distraction from their work. One member balks at the prospect of concerns with genetic modification, noting that "genetically modified organism" does not even really exist as a biological category (interview, January 14, 2018). Another voices a common community belief that genetic modification techniques are not substantially unique or novel but rather exist on a continuum with selective breeding and the mutation of plants via radiation (interview, October 17, 2018). Still, the lab members did express a willingness and form a plan to publicly debate genetic engineering, with the conviction that better understanding of the technology will lead to greater acceptance (interview, December 10, 2017). Dismissive of health and safety concerns over GMOs and glad not to have faced them too often, lab members are often more concerned with the politics of intellectual property rights and often critical of the ways these are leveraged by for-profit entities.

Although their disagreement with existing intellectual property regimes compels them to adopt a nuanced view of bioengineering practice, being forced to navigate a maze of intellectual property protections that are strictly "techno-legal" in nature (Parthasarathy 2017) takes up much of their time and limits their ability to reimagine bioengineering in practice (field notes, October 15, 2017). Although they interpret current intellectual property protections as broken—for being excessively bureaucratic and helping to foster wealth concentration and unequal distribution of life-saving drugs—they also acknowledge that any technological solution must either avoid infringing on intellectual property or face the consequences (field notes, September 27, 2017; field notes, October 15, 2017; interview, December 10, 2017). Although the biohacking or DIY-bio model certainly has its limitations, the existence of projects that encourage greater participation in and public understanding of technology-in-the-making provides a promising example of how and why to avoid fetishizing fermentation. Closer attention to diverse practices of fermentation includes a more engaged consideration of their technical elements, as well as of the social networks that support them and the differential social outcomes that even technically similar projects produce (see Latour 1987, 1988). Without such consideration, so-called fermentation (or bioengineering, or synthetic biology, for that matter) does not stand well on its own as a signifier of a particular kind of change in productive landscapes.

Conclusion: Fermentation in Theory and Practice

Borrowed concepts are necessarily used to explain new technological developments; however, bioengineers' invocation of fermentation is more than a metaphor. Biotechnological developments over the last several decades have changed the definition of fermentation itself to encompass a much broader range of culturing processes and potential outputs. Perhaps due to a general lack of public engagement with the details of what fermentation is and does, these changes are not widely known. This lack of engagement has many possible causes. In industrial society the separation between production and consumption is pervasive. While alternative food movements, for example, have brought fermentative production closer to the consumer in some forms, other forms—like the modified ferments that are the product of bioengineering—remain sequestered in laboratories and specialized literatures. Bioengineered ferments, such as insulin and synthetic rennet, are rarely sold to consumers directly. Sales of the former are heavily mediated by physicians and pharmacies and protected by intellectual property, and the latter is a product aimed at another type of fermentation producer—cheesemakers. Whatever the causes of this lack of engagement, it creates a mismatch between scientific and public understandings of fermentation. This mismatch can impede the accurate communication of technoscientific developments or, as in Perfect Day's case, be used to promote bioengineering endeavors using nominal or superficial similarities and skirting more potentially controversial elements of bioengineering practice.

Alistair Elfick and Drew Endy (2014), who are synthetic biologists, seem to lament that people do not know the benefits they receive from genetic modification, but they do not delve much into *why* these benefits are not common knowledge. Along with the separation between production and consumption, the public controversy over GMOs that has emerged from some of bioengineering's uses may make bioengineers reluctant to widely publicize the genetic engineering components of production practices, even if they believe public concerns and fears are unnecessary or misplaced. "Fermentation," without much descriptive specificity, comes to serve as a shorthand that can be used to describe a process without the need for detail and to take advantage of positive associations with more visible forms of fermentative production. However, even these more visible forms—like so-called "craft" fermentation— may be nebulous, putting practical specificity and shared understanding even further out of reach. The fraught history of science and engineering work in terms of public trust, especially when it comes to genetic modification technologies, incentivizes a superficial attempt to establish novel technology as essentially familiar and benevolent.

I have described the resulting superficial equivalence of different practices that go by the name "fermentation" as a form of fetishism. This fetishism can result in a "cooperation without consensus" (Star 1993; Clarke and Star 2008) in which more familiar fermentation practices are leveraged as convenient reference points and inoculants against public skepticism and criticism. Stakeholders in fermentation's different forms end up tacitly supporting one another without delving into how their differences—whether technical, organizational, or value-based—set them apart. As bioengineering continues to transform fermentation in profound new ways, these differences and the gaps between specialists' and publics' understandings grow larger.

My goal is to demonstrate to how putting familiar terms to new uses can yield obfuscation rather than clarity, as well as how this is already happening with fermentation. The work in this volume asks, "Does theorizing landscape change as fermentation help us understand landscape change better?" In response this chapter demonstrates some potential challenges and implications that face this line of thinking and thereby raises some further questions for consideration: If theorizing landscape change as fermentation does help us understand landscape change better, might it come at the expense of understanding fermentation better? Is this an acceptable trade-off? Given fermentation's wide variety in practice, which meaning of the word is being used to theorize landscape change? And how is this particular meaning being selected? While it is not the primary goal of this volume to better grasp fermentation itself, this risk of using fermentation as a metaphorical theoretical concept should be of concern for scholars with a specific interest in understanding fermentative industry. These scholars should be wary of the ways in which fermentation can become a nebulous signifier used to avoid difficult topics, such as the use of genetic engineering in food production.

These concerns with the use of fermentation as metaphor stem from a very specific set of developments with a tendency to play fast and loose with the concept. By exploring the merits of fermentation-as-metaphor, the present volume (see chapters 10 and 12 in particular) already demonstrates a reflexive engagement that resists some of the pitfalls of fetishizing fermentation. With any success, its contents will foster a better understanding of both fermentation and landscape change. As community biologists help demonstrate, fermentation-as-metaphor need not be off the table altogether; it is simply at its best when it takes up some of the complexity of fermentation in practice, rather than using it as convenient conceptual shorthand.

With the increasing accessibility of bioengineering technology, community biology laboratories can provide for better public engagement with both the technical and ethical dimensions of fermentative production, even though their scope and outreach are limited and they face some difficult institutional constraints. At their best these spaces adopt ways of doing bioengineering that are both explicitly political (in their aims to address problems they identify with other forms of bioengineering) and relatively accessible (to nonexperts in biology, social scientists included). Even when making analogies to other forms of fermentation, members of these laboratories do so with more care and resist some— though not all—of the issues with fetishizing fermentation. Perhaps using fermentation as a conceptual tool for theorizing landscape change can avoid these issues as well, if those doing this work remain engaged with fermentation in practice and cognizant of their own values and goals (see chapters 10, 12, and 13, this volume).

Notes

1. Many practitioners would make distinctions between the terms "bioengineering," "synthetic biology," and "metabolic engineering," but for the purposes of this chapter the important consideration is the confluence of biology and engineering as it is applied to microbial metabolisms, which can fall under any of these labels.

2. To protect the anonymity of my research participants, I have omitted their names and the names of their organizations and cited their input as "interview," along with a date, depending on whether this communication occurred during a recorded interview. Their words have in some cases been modified slightly for clarity.

3. This hints at some of the mystical associations with fermentation, a process that always has an element of the unknown and the uncertain. To a lesser extent, this association remains, as evident in longtime Anchor Brewing Company owner Fritz Maytag's attitude toward beer making: "Beer does not make itself properly by itself. It takes an element of mystery and of things that no one can understand" (White and Zainasheff 2010, xv).

4. Of course this distinction has not always been so fundamental or so clear, and biology has had repeated clashes over vitalism and basic distinctions between biotic and abiotic matter (Bud 1993).

5. Because it refutes the theory of spontaneous generation, *Studies on Fermentation* is also a key text in the development of the germ theory of disease. This theory—that sickness is caused by single-celled organisms that, rather than appearing under certain environmental conditions, are derived from previous generations of single-celled organisms—displaced miasmic and zymotic theories of disease. The latter are named after the Greek word for ferment and grew from the simultaneous (mis-) understanding of both fermentation and infectious disease as spontaneous processes (Tomes 1998).

6. "Zymotechnology" comes from the Greek word for fermentation, *zymosi*.

7. The sequencing of the genome has contributed to the increasingly prevalent understanding of life itself as an informatic domain, a complex-yet-crackable code. As for genetically modified organisms, the patentability of single-celled organisms was enabled by the ruling in *Diamond v. Chakrabarty* (SCOTUS 1980), which later paved the way for modified plants (Kloppenburg 1988) and laboratory animals with "special" features like a predisposition to developing cancer (Haraway 1997).

8. Prior to Genentech's *E. coli*–produced insulin, which the company was making by 1978 and would introduce into the marketplace a few years later, insulin derived from nonhuman animals—usually cattle ("bovine" or "beef" insulin) or pigs ("porcine" or "pork" insulin)—was the industry standard. This insulin had significant limitations, however: it was expensive to produce, it was subject to impurities, it had a tendency to provoke immune responses in patients, and unless modified, it differed in molecular composition from the insulin produced in the human pancreas. Finding new means of producing "human" insulin—that is, synthetic insulin with the same molecular structure as that produced in human bodies—was therefore a valuable achievement for the nascent field of genetic engineering (Greene and Riggs 2015).

9. When Keasling says, "We use it," the antecedent is made somewhat unclear through video editing, but it is clear from context that Keasling is referring either to yeast or to fermentation, possibly both.

10. Note the lack here of any reference to biotechnology, genetic engineering, or bioengineering.

11. Other criteria, like the limited use of machinery, are difficult to apply consistently in the contemporary moment, to craft brewing, for example.

12. "Differences that make a difference" is a reference to Bateson's definition of the word *information*.

13. Many practitioners would make a distinction or express a preference for one term or the other. For the purposes of this chapter, the terms are synonymous. Most of my fieldwork participants use both terms and use them more or less interchangeably.

References

Bateson, Gregory. 1972. *Steps to an Ecology of Mind: Collected Essays in Anthropology, Psychiatry, Evolution, and Epistemology.* Chicago: University of Chicago Press.

Bobrow-Strain, Aaron. 2012. *White Bread: A Social History of the Store-Bought Loaf.* Boston: Beacon Press.

Brewers Association. n.d. "Craft Brewer Defined." Accessed March 27, 2019. https://www.brewersassociation.org/statistics/craft-brewer-defined/.

Bud, Robert. 1993. *The Uses of Life: A History of Biotechnology.* Cambridge: Cambridge University Press.

Calvert, Jane. 2013. "Engineering Biology and Society: Reflections on Synthetic Biology." *Science, Technology and Society* 18 (3): 405–20. https://doi.org/10.1177/0971721813498501.

Calvert, Jane, and Paul Martin. 2009. "The Role of Social Scientists in Synthetic Biology." *EMBO Reports* 10 (3): 201–4. https://doi.org/10.1038/embor.2009.15.

Campbell, Colin. 2005. "The Craft Consumer: Culture, Craft and Consumption in a Postmodern Society." *Journal of Consumer Culture* 5 (1): 23–42. https://doi.org/10.1177/1469540505049843.

Clarke, Adele E., and Susan Leigh Star. 2008. "The Social Worlds Framework: A Theory/Methods Package." In *The Handbook of Science and Technology Studies*, edited by Edward J. Hackett, Olga Amsterdamska, Michael Lynch, and Judy Wajcman, 113–37. 3rd ed. Cambridge MA: MIT Press.

CNN. 2013. "Jay Keasling: Using Microbes to Create the Next Generation of Fuel." CNN.com, February 5, 2013. http://www.cnn.com/2013/02/05/tech/jay-keasling-using-microbes-to-create-the-next-generation-of-fuel/index.html.

Cohen, Stanley, and Herbert Boyer. 1980. Process for producing biologically functional molecular chimeras. U.S. Patent 4237224A, filed January 4, 1979, and issued December 2, 1980. https://patents.google.com/patent/us4237224a/en?q=insulin&inventor=herbert+boyer.

DuPuis, E. Melanie. 2015. *Dangerous Digestion: The Politics of American Dietary Advice.* Oakland: University of California Press.

Elfick, Alistair, and Drew Endy. 2014. "Synthetic Biology: What It Is and Why It Matters." In *Synthetic Aesthetics: Investigating Synthetic Biology's Designs on Nature*, edited by Alistair Elfick and Drew Endy, 3–26. Cambridge MA: MIT Press.

El-Mansi, E. M. T., ed. 2012. *Fermentation Microbiology and Biotechnology.* 3rd ed. Boca Raton: Taylor & Francis/CRC Press.

Endy, Drew, Stephanie Galanie, and Christina D. Smolke. 2015. "Complete Absence of Thebaine Biosynthesis under Home-Brew Fermentation Conditions." *bioRxiv*, posted August 13, 2015. https://doi.org/10.1101/024299.

Foo, Jee Loon, Heather M. Jensen, Robert H. Dahl, Kevin George, Jay D. Keasling, Taek Soon Lee, Susanna Leong, and Aindrila Mukhopadhyay. 2014. "Improving Microbial Biogasoline Production in *Escherichia coli* Using Tolerance Engineering." *mBio* 5 (6): e01932–14. https://doi.org/10.1128/mBio.01932-14.

Franklin, Sarah. 2007. *Dolly Mixtures: The Remaking of Genealogy.* Durham: Duke University Press.

Galanie, Stephanie, Kate Thodey, Isis J. Trenchard, Maria Filsinger Interrante, and Christina D. Smolke. 2015. "Complete Biosynthesis of Opioids in Yeast." *Science* 349 (6252): 1095–1100. https://doi.org/10.1126/science.aac9373.

Genentech. 1978. "First Successful Laboratory Production of Human Insulin Announced." Genentech Press Releases, September 6, 1978. https://www.gene.com/media/press-releases/4160/1978-09-06/first-successful-laboratory-production-o.

Greene, Jeremy A., and Kevin R. Riggs. 2015. "Why Is There No Generic Insulin? Historical Origins of a Modern Problem." *New England Journal of Medicine* 372 (12): 1171–75. https://doi.org/10.1056/NEJMms1411398.

Hale, Victoria, Jay D. Keasling, Neil Renninger, and Thierry T. Diagana. 2007. "Microbially Derived Artemisinin: A Biotechnology Solution to the Global Problem of Access to Affordable Antimalarial Drugs." In "Defining and Defeating the Intolerable Burden of Malaria III: Progress and Perspectives," edited by J. G. Breman, M. S. Alilio, and N. J. White. Supplement, *American Journal of Tropical Medicine and Hygiene* 77 (6): 198–202. https://doi.org/10.4269/ajtmh.2007.77.198.

Haraway, Donna Jeanne. 1997. *Modest_Witness@Second_Millennium: FemaleMan_Meets_OncoMouse; Feminism and Technoscience*. New York: Routledge.

Hayden, Erika. 2014. "Synthetic-Biology Firms Shift Focus." *Nature News* 505 (7485): 598. https://doi.org/10.1038/505598a.

Hayles, N. Katherine. 1999. *How We Became Posthuman: Virtual Bodies in Cybernetics, Literature, and Informatics*. Chicago: University of Chicago Press.

Jasanoff, Sheila. 2005. *Designs on Nature: Science and Democracy in Europe and the United States*. Princeton: Princeton University Press.

Keller, Evelyn Fox. 2002. *Making Sense of Life: Explaining Biological Development with Models, Metaphors, and Machines*. Cambridge MA: Harvard University Press.

Kloppenburg, Jack Ralph. 1988. *First the Seed: The Political Economy of Plant Biotechnology*. Madison: University of Wisconsin Press.

Landecker, Hannah. 2007. *Culturing Life: How Cells Became Technologies*. Cambridge MA: Harvard University Press.

Latour, Bruno. 1987. *Science in Action: How to Follow Scientists and Engineers through Society*. Cambridge MA: Harvard University Press.

———. 1988. *The Pasteurization of France*. Cambridge MA: Harvard University Press.

Liu, Yanfeng, Hyun-dong Shin, Jianghua Li, and Long Liu. 2015. "Toward Metabolic Engineering in the Context of System Biology and Synthetic Biology: Advances and Prospects." *Applied Microbiology and Biotechnology* 99 (3): 1109–18. https://doi.org/10.1007/s00253-014-6298-y.

Marx, Karl. 2004. *Capital: A Critique of Political Economy*. N.p.: Penguin UK.

Murray, Andy. 2018. "Meat Cultures: Lab-Grown Meat and the Politics of Contamination." *BioSocieties* 13 (2): 513–34. https://doi.org/10.1057/s41292-017-0082-z.

National Academies of Sciences, Engineering, and Medicine. 2016. *Genetically Engineered Crops: Experiences and Prospects*. Washington DC: National Academies Press.

Parthasarathy, Shobita. 2017. *Patent Politics: Life Forms, Markets, and the Public Interest in the United States and Europe*. Chicago: University of Chicago Press.

Pasteur, Louis. 1879. *Studies on Fermentation: The Diseases of Beer, Their Causes, and the Means of Preventing Them*. London: Macmillan.

Pauly, Philip J. 1987. *Controlling Life: Jacques Loeb and the Engineering Ideal in Biology*. New York: Oxford University Press.

Paxson, Heather. 2012. *The Life of Cheese: Crafting Food and Value in America*. Berkeley: University of California Press.

Perfect Day. 2016a. "FAQs." http://www.perfectdayfoods.com/faq/.

———. 2016b. "Perfect Day: All the Dairy You Love, with None of the Dairy Cows." http://www.perfectdayfoods.com/.

Rabinow, Paul. 1996. *Making PCR: A Story of Biotechnology*. Chicago: University of Chicago Press.

Rabinow, Paul, and Gaymon Bennett. 2012. *Designing Human Practices: An Experiment with Synthetic Biology*. Chicago: University of Chicago Press.

Ro, Dae-Kyun, Eric M. Paradise, Mario Ouellet, Karl J. Fisher, Karyn L. Newman, John M. Ndungu, Kimberly A. Ho, et al. 2006. "Production of the Antimalarial Drug Precursor Artemisinic Acid in Engineered Yeast." *Nature* 440 (7086): nature04640. https://doi.org/10.1038/nature04640.

Roosth, Sophia. 2017. *Synthetic: How Life Got Made*. Chicago: University of Chicago Press.

Schurman, Rachel A., and Dennis D. Kelso. 2003. *Engineering Trouble: Biotechnology and Its Discontents*. Berkeley: University of California Press.

SCOTUS (Supreme Court of the United States). 1980. Diamond v. Chakrabarty, 447 U.S. 303. Accessed January 28, 2018. https://supreme.justia.com/cases/federal/us/447/303/.

Shiva, Vandana. 1997. *Biopiracy: The Plunder of Nature and Knowledge*. Boston: South End Press.

Shotwell, Alexis. 2016. *Against Purity: Living Ethically in Compromised Times*. Minneapolis: University of Minnesota Press.

Smolke, Christina, ed. 2009. *The Metabolic Pathway Engineering Handbook: Tools and Applications*. Boca Raton FL: CRC Press. https://books.google.com/books/about/The_Metabolic_Pathway_Engineering_Handbo.html?id=wzAPgK1ulBQC.

Stanbury, Peter F., Allan Whitaker, and Stephen J. Hall. 2014. *Principles of Fermentation Technology*. 2nd ed. Kent, England: Elsevier Science.

Star, S. L. 1993. "Cooperation without Consensus in Scientific Problem Solving: Dynamics of Closure in Open Systems." In *CSCW: Cooperation or Conflict?*, edited by Steve Easterbrook, 93–106. London: Springer. https://doi.org/10.1007/978-1-4471-1981-4_3.

Steen, Eric J., Yisheng Kang, Gregory Bokinsky, Zhihao Hu, Andreas Schirmer, Amy McClure, Stephen B. del Cardayre, and Jay D. Keasling. 2010. "Microbial Production of Fatty-Acid-Derived Fuels and Chemicals from Plant Biomass." *Nature* 463 (7280): 559–62. https://doi.org/10.1038/nature08721.

Stephanopoulos, George, Aristos A. Aristidou, and Jens Nielsen. 1998. *Metabolic Engineering: Principles and Methodologies*. San Diego: Academic Press.

Thodey, Kate, Stephanie Galanie, and Christina D. Smolke. 2014. "A Microbial Biomanufacturing Platform for Natural and Semisynthetic Opioids." *Nature Chemical Biology* 10 (10): 837–44. https://doi.org/10.1038/nchembio.1613.

Thompson, Charis. 2005. *Making Parents: The Ontological Choreography of Reproductive Technologies*. Cambridge MA: MIT Press.

Tomes, Nancy. 1998. *The Gospel of Germs: Men, Women, and the Microbe in American Life*. Cambridge MA: Harvard University Press.

White, Chris, and Jamil Zainasheff. 2010. *Yeast: The Practical Guide to Beer Fermentation*. Boulder CO: Brewers Publications.

Raw Power   12

For a (Micro)biopolitical Ecology of Fermentation

Eric Sarmiento

Fermented foods and beverages have received considerable popular and scholarly attention in recent years, especially in the Global North. Food celebrities such as Michael Pollan and Sandor Katz give public lectures around the United States and abroad and write extensively on fermentation, and the markets for fermented products ranging from craft beers and kombucha to kimchi and yogurt have expanded dramatically. These trends are arguably part of a broader cultural shift in the United States, in particular toward gourmet and specialty food culture and popular explorations of a range of "alternative" food products (e.g., organics and local foods) and initiatives (e.g., community-supported agriculture and fisheries, community gardens, and fair trade). Fermented foods articulate with this broader shift in several ways: they are valued for their culinary qualities, adding tastes and textures to dishes; they are celebrated as integral components of many traditional food cultures; and they are increasingly viewed as an important and perhaps essential element in a healthy diet. In all three of these aspects the quality or value of fermented foods and beverages is tightly bound up with the microbial actors who, through fermentation, transform the taste, texture, and biochemical capacities of food and drink.

In this chapter I focus on the ways in which the actions and capacities of microbes are intertwined with political struggles embedded in the production and consumption of fermented foods and beverages and their associated landscapes. The goal of the chapter is to briefly review several strands of thought emerging in scholarly and popular discussions of fermentation, the human microbiome, food systems, and health. In synthesizing these lines of thinking, I draw attention to some politically charged features of fermented landscapes that would benefit from further study. I argue that future research should emphasize a political ecological approach to studying fermentation but more specifically a political ecology in which the political includes the biopolitical and ecology is understood not as an environment or surrounding milieu but as a continuous material fabric that is not only around us but also on and within human bodies. This conception of fermentation and its associated landscapes highlights the importance of distinguishing between pasteurized fermented food and drink and unpasteurized or "raw" ferments, as pasteurization eradicates the microbial partners that are increasingly understood as essential to the internal ecologies of our bodies. We shall see that for many advocates of raw fermented foods and beverages the presence or absence of microbes matters in terms of not only individual and public health but also the situated, power-laden political ecologies of place.

The chapter is structured as follows. In the next section I offer a sketch of the history of human-microbial relations in the Global North over the past century or more, focusing on recent transformations in our understanding of microbes as essential to human health. This historical sketch provides context for exploring how food health and safety regulations mediate the markets for fermented foods and beverages. In the subsequent section I focus on the politics of health and illness, touching on the spatial distribution of fermentation and examining debates about how emerging expert knowledge of the microbiome might be mobilized and in whose interests. From there I consider how attitudes and subject positions vis-à-vis microbes are an important ground of struggle for the political ecologies of raw ferments. As politics far exceed the choices and subjective decisions of individuals, however, I explore in the penultimate section the post-Pasteurian argument that raw ferments are a form of resistance to the homogenizing influence of capitalist globalization. In the final section I recapitulate the discussion and offer some concluding remarks.

Microbes, Fermentation, and the
Regulatory Biopolitics of Food Systems

Microbial activity has been a matter of concern in food production, distribution, and consumption since (at least) the late nineteenth century, a period when microbiologists made major strides in understanding the role of microorganisms in spreading disease (Latour 1988). The Pasteurian regime that emerged from this period framed microbes as an enemy and rested more broadly on a narrative of humans being at war with microbes in the food system (Latour 1988). Food supply in the Global North increased greatly during this period, and food-related illnesses declined markedly in densely crowded industrial cities fractured by stark socioeconomic inequalities. This paradigm shift in contending with microbial activity, however, should be understood as an adjunct to the extension of industrial production models to food systems that was occurring at the same time. During this period food production became increasingly mechanized and intensive, and farms as well as distribution and retail operations were consolidated and expanded dramatically in scale (Goodman, Wilkinson, and Sorj 1987). Such industrialized production sites concentrate high volumes of food products and production processes, potentially operating as breeding grounds for disease and microbial proliferation and increasing the likelihood that contaminants will come into contact with uncontaminated items. These sites are also typically components of far-reaching distribution chains, thus expanding the potential geographic reach of contamination and outbreaks.

The health and safety concerns that emerged from the industrial restructuring of food systems contributed to the sense of warring with microbes, a food safety regime that the food historian Felipe Fernández-Armesto (2001) refers to as a quest for "purity" in food science. As several scholars have pointed out (Kurtz et al. 2013; Spackman 2017; Speake 2011), this Pasteurian approach exemplifies what Michel Foucault (2007, 2008) has referred to as biopolitical governance, in that policy measures seek to securitize the health of the population while maximizing the productive capacities of the national territory and governing by managing the intersection of biological, social, and economic dynamics (see also Braun 2000; and Rabinow and Rose 2006). In broad terms the biopolitical public health initiatives that followed in the wake of the germ theory of disease and the industrialization of agriculture were mutually enabling phenomena, ushering in a relatively stable discursive regime that underpinned the theory and practice of food health and safety in much of the Global North for the majority of the twentieth century.

One implication of that Pasteurian regime is that purity-oriented food safety regulations have profoundly shaped food markets, as these regulations are generally geared toward large-scale industrial production and distribution systems, which require considerable financial investment and raise per-unit costs. As such, this approach presents significant barriers to small-scale producers and thus has contributed to the consolidation and concentration of food production into fewer, more industrialized firms (DeLind and Howard 2008; Dupuis 2002; Hassanein 2011; Sarmiento 2015b). Of particular interest to the fermented landscapes research agenda, however, is that scientific and popular understandings of microbes have undergone a profound transformation in recent years, as both experts and food consumers begin not only to distinguish between harmful and beneficial microbial actors but also to reimagine the relationships among microorganisms, ecology, human bodies, and landscapes, with profound implications for the politics of food safety regulations.

At the heart of this transformation in our understanding of microbes is a plethora of recent research on the human microbiome. This work has centered on the massive research endeavor known as the Human Microbiome Project (HMP), an international consortium involving approximately 200 scientists working at 80 institutions around the world to catalog, categorize, and sequence the genetic material of the microbial communities that live in and on human bodies (Human Microbiome Project 2018). These communities, it turns out, are both extensive and astonishingly diverse: the average healthy human adult is home to hundreds of trillions of bacteria, and microbial cells in and on our bodies outnumber our human cells by a considerable ratio. Moreover, these commensal microbial communities are vastly more diverse in genetic terms than our own human genome and appear to be capable of mutating much more easily than human genes. These findings have generated a colossal amount of published research—the National Institutes of Health report that by the end of 2017, more than 650 papers on this topic had been published by HMP researchers alone and had already been cited more than 70,000 times—with ongoing and new research initiatives proliferating (Human Microbiome Project 2018).

Microbiome research has profoundly affected thinking in many disciplines, from public health and immunology to cultural anthropology and geography, as scholars grapple with the far-ranging implications of our growing awareness that in some respects we are in fact at least as much microbe as human. The microbiome is linked to our metabolic function, immune systems, and even mood and cognition; a range of diseases and conditions, from autoimmune diseases to depression, thus appear to correlate with disturbed or compromised microbial ecologies in the body (Foster and McVey Neufeld 2013; Jandhyala et al. 2015). Given this existential interdependence with bacteria and other components of the microbiome, we are called upon to rethink the essence of what it means to be human and question the ontological status of the individual organism. A new language is emerging in response to this challenge: a full accounting of the genetic resources that constitute humans requires consideration of both the human genome and that of our bacterial fellow travelers, which together constitute a "metagenome," wherein human individuals are perhaps best thought of as "holobionts," or "wholes that emerge in community with a host of others" (Schneider and Winslow 2014). In this framing of human life, we emerge and develop in conjunction with the microbial worlds within us, as well as around us (more on this below), coexisting in a complex and continuous ecological fabric, much of which is invisible to the eye. This fabric covers us, extends deep into our bodies, and surrounds us. For *Homo microbis*

(Helmreich 2014), there are no clear boundaries between inside and outside, and our agencies are clearly not ours alone.

Such a radically different understanding of human-microbe relations challenges and complicates the long-standing Pasteurian approach to food safety, with important implications for fermented foods and beverages. As it becomes increasingly clear that an all-out war-like stance toward microbes may have serious negative repercussions for human health, the quest for purity begins to give way to a more ecological approach to food safety, one in which human food practices must work *with* rather than against microbes. Social science research in geography, anthropology, and other fields has begun to explore the cultural and political ecological dynamics of this rupture in food safety and health and nutrition discourse. As an example, Heather Paxson's (2008) ethnographic work with raw-milk cheese producers and consumers demonstrates that many of these individuals question the foundational assumptions of the dominant Pasteurian approach to food safety, arguing instead that we should view the human-microbe relation as collaborative, as contingent on political ecological practices, and as potentially symbiotic, rather than always as simply pathological (see also Lorimer 2016, 2017). These "post-Pasteurians" contend that for most people, raw-milk cheese and other fermented foods may well be safer to eat than pasteurized foods, "for what protects the cheese protects us" (Paxson 2008, 32). From this perspective, the diverse ecosystems and beneficial microbes and metabolites present in raw-milk cheeses inhibit colonization by harmful microbes, whereas the sterilized ecosystems presented by cheeses made from pasteurized milk provide a less competitive environment in which harmful microbes can proliferate in the event of contamination.

However, while such proponents of raw fermented food and drink may be post-Pasteurian, the regulatory landscapes in which they operate are still largely focused on waging war against microbes in food systems, which presents challenges for producers and consumers of raw ferments. Critics of conventional food systems argue that this is precisely the point of upholding Pasteurian standards—to maintain market dominance by large, industrial firms. As Paxson notes (2008, 32), "the National Cheese Institute, the dairy association whose 90 members collectively manufacture about 80 percent of cheese, processed cheese, and cheese products in the United States, is reportedly lobbying for mandatory pasteurization of all cheese milk." This move would eliminate from the market all raw-milk cheese producers, the majority of which are small-scale operations selling their products locally and regionally. Paxson and other critical scholars (Ingram 2007; Kurtz, Trauger, and Passidomo 2013) have made important strides in tracing how health and safety regulations pertaining to microbial activity can serve as an avenue through which competing approaches to food systems seek to shape markets for unpasteurized, or "lively," foods, but this is an area of research that calls for further study as post-Pasteurian ways of relating to food become more widespread. Recent controversies surrounding the regulation of kombucha, for example, warrant closer critical scrutiny. The U.S. market for kombucha, perhaps the most ubiquitous raw (though, now, in its more commercial forms, sometimes pasteurized) fermented beverage, has dramatically expanded in recent years (see chapter 9, this volume), and if PepsiCo's purchase of major brand KeVita is any indicator, food industry giants are acting to take advantage of this expansion as part of a broader move toward the marketing of "functional foods" (Troitino 2017). It seems likely that regulations will mediate this growing market as kombucha continues the rapid shift from small-scale artisanal production to mass-produced corporate commodity.

The political ecological dynamics of emerging understandings of microbes and fermentation, however, extend beyond regulatory politics, with implications for health and medicine, food subjectivities, and control over the production of place. In the following sections I consider each of these threads in turn.

Fermentation and the Politics of Health and Medicine

As noted above, emerging knowledge of the human microbiome increasingly emphasizes that healthy microbial ecologies in and around our bodies are essential to good health and well-being. Proponents of raw fermented foods and beverages have been arguing for decades that the hermetic, sterilized nature of industrialized diets is a major part of what makes these diets unhealthy (Fallon and Enig 1999; Katz 2016; Mintz 2011), given that the probiotic nature of raw ferments has historically for most human cultures played a crucial functional role in increasing the bioavailability of nutrients in food and maintaining a healthy gut microbiome.[1] This raises important questions about the social and spatial distribution of access to raw ferments and of probiotic and pathological microbial ecologies more broadly. It appears increasingly likely, for instance, that the absence of raw fermented foods and beverages in industrialized, Pasteurian food systems could be contributing to a range of health conditions that plague much of the Global North, from chronic digestive disorders to issues with mood and cognition. Conversely, as Sidney Mintz (2011) points out, probably more than half of the food fermentation occurring globally takes place in the Global South, in places where Pasteurianism and conventional agribusiness are not as extensively adopted. If people living in these cultural milieus benefit from coexisting with beneficial microbes, then a crucial direction for research in the Global South will be to explore how human-microbial relations change alongside the introduction of conventional agribusiness methods, rationales, and relations of production and consumption. Put a bit differently, cultures that have not fully adopted Pasteurian and industrial food practices likely have much to teach those of us living amid conventional food systems (Rueda Esquibel and Calvo 2013).

That said, a rejection of Pasteurianism writ large based on a romanticized notion of traditionality would be a mistake. Informal urban settlements, for instance, many of which are not served by water and sewage infrastructure, likely present microbial ecologies that are dramatically different from those of a gated community in the same city, to say nothing of a suburban home in the Global North where people may be experimenting with raw ferments. Moreover, the divergent human-microbial ecologies of these places are enmeshed in far-reaching assemblages mediated by power dynamics operating at a range of scales, from local and regional development priorities to global investments in real estate and infrastructure. Raw ferments and the microorganisms that help to produce them take on meaning and capacities in conjunction with the diverse actors associated with such assemblages (see also chapter 10, on knowledge production and fermentation, and chapter 11, on zymurgeography).

Such a relational understanding of fermentation draws our attention to several problems with—and threats to—growing scientific and public interest in microbes and fermentation. First, the emergence of the microbiome as an object of knowledge is already entangled with the efforts of various actors to capture economic value through commoditizing the therapeutic capacities or functions of microbial actors. Melody Slashinski et al. (2012) interviewed sixty-three microbiome researchers and found that many of them were concerned about the proliferation of probiotic capsules and other products that claim to provide health benefits but are not regulated as stringently as pharmaceutical drugs. Curiously, however, their study does not report concerns among these researchers about the ways in which pharmaceutical companies are also seeking to capture "biovalue" from microbial capacities. Jamie Lorimer (2017, 553) finds, for instance, that scientists decoding the hookworm genome envision a "'veritable pharmacopeia' of synthetic molecules that will become available for new drug development," a position that appears to mirror the reductionist, exchange value–capturing logic of globalized industrial food systems. This approach has come under fire from other scientists who point to "the differences between a drug that is designed to treat a specific illness and a live organism that acts preventatively on a bodily ecology to confer health[, suggesting] that a focus on drugs from bugs ignores the systemic drivers of dysbiosis" (553).

Raw ferments, as noted above with the case of kombucha, are similarly subject to commodification and control by corporate actors. But unlike hookworm therapy, many fermented foods and beverages are remarkably easy to make at home, requiring no special equipment and only basic knowledge that can be readily accessed online, from a range of widely available popular texts, or learned from communities of practitioners (see chapter 9, this volume). Additionally, while commercial unpasteurized fermented foods and beverages do tend to be expensive specialty items, home fermentation is very inexpensive in general. These low barriers to access have led Sandor Katz, Sally Fallon, and other post-Pasteurians to celebrate these raw ferments as democratic and empowering, helping people in the Global South to defend their traditional food cultures and approaches to health and nutrition and assisting people in the Global North in taking back their food systems and their own personal health. Questions remain, however, regarding cultural attitudes toward raw ferments and microbes, particularly in places where Pasteurianism has long influenced food subjectivities, a topic to which I now turn.

Disciplining Lively Food Subjects

Paxson, Lorimer, and others have pointed out that post-Pasteurianism connotes a politically salient shift in subjectivity vis-à-vis food and microbes: in Paxson's (2008) terms, Pasteurian approaches to food safety and food culture have over several generations cultivated "germophobic" subjects who are convinced that microbes are the enemy, who prefer convenience over nourishment, and whose tastes have been acclimated to microbially "dead" industrial foods (see also Carolan 2011; and Katz 2006). This observation calls into question the ease with which some consumers and producers might incorporate raw ferments into dietary practices or livelihood strategies. Such embodied sensibilities toward food and microbes are thus viewed as a critical terrain of contestation over the future of food systems and individual and collective health, as various actors seek to influence the subjective stances of producers and consumers toward raw fermented foods and beverages. In a Foucaultian approach, processes of subjection formation, of being *disciplined* as particular kinds of subjects, are not something to be liberated from but are rather structuring forces necessary to life (Foucault 2007, 2008). Thus, a fundamental question for what Foucault referred to as "the ethical cultivation of the self" (Foucault 1988, part two) is not how to avoid or escape from discipline but rather who or what disciplines us, as well as how we might be disciplined—or discipline ourselves—differently.

Producing and consuming raw fermented foods presents a *potential* avenue through which to cultivate a post-Pasteurian subjectivity, to "learn to be affected" (Latour 2004) by our microbial partners and the complex, dense webs of ecological relations traversing our bodies, our surroundings, the foods we ingest, and the environments that those foods move through from their places of production to our mouths and beyond. As Paxson (2008, 40) puts it, "a post-Pasteurian care of the self goes through the obligatory passage point of caring for the microbe—the good microbe, the *Lactobacillus* or *Penicillium* companion species whose bodies and cultures are coproduced with human ones." Exploring the complex web of actors that influence such subjective transformations is another important area for future research on fermented landscapes. For example, chronic illnesses of the sort that are increasingly common in the Global North could potentially destabilize the dominant discourses underpinning Pasteurian subjectivities (Sarmiento 2015a), particularly if emerging expert knowledge of the microbiome also undermines Pasteurianism as it links chronic illness with compromised microbial ecologies of the body. As with the politics of food health and safety regulations, critical scrutiny of competing discourses among researchers and scholarly disciplines will thus be an integral factor in understanding how subjectivities vis-à-vis lively ferments take shape moving forward.

Influences on food subjectivities also include class, race, and gender dynamics, among other factors. The alternative food networks (AFNs) with which raw ferments are frequently linked in the Global North have been extensively critiqued as exclusive, primarily white and privileged milieus (Guthman 2003; Slocum 2007). Moreover, as critical scholars have noted, individual production and consumption choices alone are not enough to effect structural changes in food systems, given profound inequities in food access and the power of corporate actors to influence policy and regulations (Johnston 2008). In this context, combating the negative impacts of Pasteurianism and effecting meaningful transformations of food systems writ large will require the cultivation of subjects who are not just microbiophilic but also committed to addressing structural inequalities and injustice in the food system. With that point in mind, I turn now to consider the democratizing potential of experimenting with raw ferments, particularly as these foods and beverages intersect with notions of food sovereignty.

Landscapes of Lively Food Sovereignty

As AFNs seek to address the problem of exclusivity and "foodie elitism," the concept of food sovereignty has begun to inform some AFN discourse and practice, particularly in urban areas where AFN initiatives intersect with processes such as food desertification and gentrification (Alkon and Mares 2012; Block et al. 2012). Emerging from land reform movements in the Global South such as Via Campesina in Latin America, the concept of food sovereignty asserts "that all people have the right 'to healthy and culturally appropriate food produced through ecologically sound and sustainable methods, and their right to define their own food and agriculture systems'" (Via Campesina 2009, quoted in Alkon and Mares 2012, 347).

Some post-Pasteurians frame raw ferments in terms that begin to resonate with food sovereignty discourse. For example, regulatory interventions that seem to favor industrial agribusiness over small-scale producers, as discussed above, are, in the eyes of post-Pasteurians, simply another form of corporate control of the food system at the expense of people's right to decide for themselves how they want their food produced and distributed. These ideas are also tied to cultural appropriateness and cultural geographical difference—regulating small, artisanal producers out of existence looks to some like another mechanism through which capitalist globalization replaces cultural difference with standardized commodities that have no apparent connection to particular places or traditions (Phillips 2006). Producing and consuming raw ferments are then viewed as acts of political resistance to the homogenizing and colonizing tendencies of global capital. As Katz (2016, 24) puts it, "Wild fermentation is the opposite of homogenization and uniformity, a small antidote you can undertake in your home, cultivating broad communities of organisms indigenous to your food, and also contributing those of your hands and your kitchen, to produce unique fermented foods. What you ferment with the organisms around you is a manifestation of your specific environment, and it will always be a little different." Paxson (2008) similarly suggests that we view raw-milk cheeses and the cultivated microbial transformations that constitute them as "nature-culture hybrids" (25), as expressions of people's connections to pieces of land, and as a "biotechnology for regionalism" (26). Seen in this light, producing and consuming raw ferments are biopolitical acts, an element of reclaiming control of what one eats and how one relates to the landscapes of food production, while also defending cultural heterogeneity from globalized, homogenizing, capitalist agrifood corporations. The future of fermented foods and drinks will, however, continue to feature the struggle between these democratic aspirations and the capitalist imperative to co-opt, privatize, and capture biovalue from microbial actors (see chapters 10 and 11 for more on this).

Note here that once again rawness is fundamental: for ferments to serve as an "antidote" to the cultural imperialism of globalized capital and agribusiness in the manner intended by Katz and other post-Pasteurians, those ferments must remain woven into the extensive/intensive ecological fabric of the landscapes in which they are embedded. Rawness, for post-Pasteurians, preserves those mutualistic threads that bind ferments and bodies within a landscape, whereas pasteurization cuts those threads, standardizes a product, enables its insertion into Pasteurian food safety regimes, and facilitates mass production and global distribution.

Rawness, however, is clearly not sufficient to bind fermentation and its associated landscapes to food sovereignty or to usher in broader and deeper changes to food systems. A salient concern for fermented landscapes research is to explore how raw ferments as technologies for re-embedding foodways in particular places might also help to create solidarities across peoples with social, cultural, and political economic differences. This is necessary for avoiding defensive localism (Winter 2003) and the more reactionary strands of political thought and practice that have emerged in response to capitalist globalization (Williams 2017).

Conclusion

In this chapter I have offered a brief overview of some of the ways that fermented landscapes are landscapes of political ecological struggle characterized by emergent and complex relations between differently situated people, microbes, and the extensive *and* intensive ecological fabric that weaves together these and other actors. At the heart of these struggles is a rapidly changing understanding of human-microbial relations. Ongoing research into the microbiome has by many accounts the potential to revolutionize our understanding of health and disease by reconfiguring the ontological status of individuals and their relationship to their particular environment and by enabling us to understand our own bodies as parts of extensive and intensive ecological tapestries. Scholars in fields ranging from geography and anthropology to epidemiology and literary studies have taken a keen interest in these potentially profound cultural shifts. If this potential develops as these scholars expect, it would seem likely that post-Pasteurian stances toward microbes, diet, landscape, place, and self could become more widespread as people learn about the importance of cultivating symbiotic relations with microbes in and around our bodies.

Such transformations in our understanding of human-microbial relations raise an issue with how we use the term "fermented": if indeed there are stark differences in the health implications of consuming pasteurized and unpasteurized fermented foods and beverages, it would seem problematic to consider all fermented foods and beverages to constitute a single category (see also chapter 11). Marking the distinction between raw and pasteurized ferments, however, is not sufficient for understanding the salience of fermented landscapes in terms of individual and public health, food justice, or food sovereignty. An effective and politically aware research agenda must interrogate and critique power relations entangled with raw ferments in the ways I have outlined throughout the chapter. Beyond critique, there are also opportunities to conduct participatory and community-based research projects that actively seek to work with situated communities in cultivating individual and collective political will to effect broader transformations in food systems.

Effective efforts to both critique power configurations in fermented landscapes and actively construct more just, democratic food systems must also take into account that the political ecologies of fermentation are partly *biopolitical*, in the Foucaultian sense of the term; we see in food safety regulations and increasingly in the world of biomedicine an attempt to govern nature/society relations by managing populations of people and microbes in an effort to secure both public health and economic value, which are intertwined. Such biopolitical ecologies of fermentation and health arguably bring a new dimension to Foucault's notion of *micropolitics*, expanding the "capillaries" of power relations to the microscopic realm and revealing that the ongoing cultivation of subjectivity is profoundly linked to our relations with microorganisms with the capacity to shape our mood, cognition, and overall health in ways that we are only beginning to grasp. Environmental degradation can thus occur not just in the landscapes around us but also within our bodies. Some of our allies in the struggle against such degradation are microbial.

Notes

1. But see the article by Albenberg and Wu (2014), whose review makes clear that the connections between diet and the intestinal microbiome are immensely complex and remain largely elusive for gastroenterological research.

References

Albenberg, Lindsey, and Gary Wu. 2014. "Diet and the Intestinal Microbiome: Associations, Functions, and Implications for Health and Disease." *Gastroenterology* 146 (6): 1564–72.

Alkon, Alison Hope, and Teresa Marie Mares. 2012. "Food Sovereignty in US Food Movements: Radical Visions and Neoliberal Constraints." *Agriculture and Human Values* 29 (3): 347–59.

Block, Daniel, Noel Chávez, Erika Allen, and Dinah Ramirez. 2012. "Food Sovereignty, Urban Food Access, and Food Activism: Contemplating the Connections through Examples from Chicago." *Agriculture and Human Values* 29 (2): 203–15.

Braun, Bruce. 2000. "Producing Vertical Territory: Geology and Governmentality in Late Victorian Canada." *Cultural Geographies* 7 (1): 7–46.

Carolan, Michael. 2011. *Embodied Food Politics*. Burlington VT: Ashgate.

DeLind, Laura, and Philip Howard. 2008. "Safe at Any Scale? Food Scares, Food Regulation, and Scaled Alternatives." *Agriculture and Human Values* 25 (3): 301–17.

DuPuis, E. Melanie. 2002. *Nature's Perfect Food: How Milk Became America's Drink*. New York: New York University Press.

Fallon, Sally, and Mary Enig. 1999. *Nourishing Traditions: The Cookbook That Challenges Politically Correct Nutrition and the Diet Dictocrats*. Washington DC: New Trends.

Fernández-Armesto, Felipe. 2001. *Food: A History*. London: Macmillan.

Foster, Jane, and Karen-Anne McVey Neufeld. 2013. "Gut-Brain Axis: How the Microbiome Influences Anxiety and Depression." *Trends in Neurosciences* 36 (5): 305–12.

Foucault, Michel. 1988. *The History of Sexuality: Volume 3, The Care of the Self*. Translated by Robert Hurley. New York: Vintage.

———. 2007. *Security, Territory, Population: Lectures at the Collège de France 1977–1978*. Translated by Graham Burchell. New York: Palgrave Macmillan.

———. 2008. *The Birth of Biopolitics: Lectures at the Collège de France 1978–1979*. Translated by Graham Burchell. New York: Palgrave Macmillan.

Goodman, David, John Wilkinson, and Bernardo Sorj. 1987. *From Farming to Biotechnology: A Theory of Agri-Industrial Development*. New York: Blackwell.

Guthman, Julie. 2003. "Fast Food/Organic Food: Reflexive Tastes and the Making of 'Yuppie Chow.'" *Social & Cultural Geography* 4 (1): 45–58.

Hassanein, Neva. 2011. "Matters of Scale and the Politics of the Food Safety Modernization Act." *Agriculture and Human Values* 28 (4): 577–81.

Helmreich, Stefan. 2014. "*Homo microbis*: The Human Microbiome, Figural, Literal, Political." *Thresholds* (42): 52–59.

Human Microbiome Project. 2018. "Program Snapshot." National Institutes of Health Office of Strategic Coordination—The Common Fund, May 2018. https://commonfund.nih.gov/hmp.

Ingram, Mrill. 2007. "Disciplining Microbes in the Implementation of US Federal Organic Standards." *Environment and Planning A* 39 (12): 2866–82.

Jandhyala, Sai Manasa, Rupjyoti Talukdar, Chivkula Subramanyam, Harish Vuyyuru, Mitnala Sasikala, and D. Nageshwar Reddy. 2015. "Role of the Normal Gut Microbiota." *World Journal of Gastroenterology* 21 (29): 8787–803.

Johnston, Josée. 2008. "The Citizen-Consumer Hybrid: Ideological Tensions and the Case of Whole Foods Market." *Theory and Society* 37 (3): 229–70.

Katz, Sandor Ellix. 2006. *The Revolution Will Not Be Microwaved*. White River Junction VT: Chelsea Green.

———. 2012. *The Art of Fermentation*. White River Junction VT: Chelsea Green.

———. 2016. *Wild Fermentation: The Flavor, Nutrition, and Craft of Live-Culture Foods*. Rev. ed. White River Junction VT: Chelsea Green.

Kurtz, Hilda, Amy Trauger, and Catarina Passidomo. 2013. "The Contested Terrain of Biological Citizenship in the Seizure of Raw Milk in Athens, Georgia." *Geoforum* 48 (August): 136–44.

Latour, Bruno. 1988. *The Pasteurization of France*. Translated by Alan Sheridan and John Law. Cambridge MA: Harvard University Press.

———. 2004. "How to Talk about the Body? The Normative Dimensions of Science Studies." *Body & Society* 10 (2–3): 205–29.

Lorimer, Jamie. 2016. "Gut Buddies: Multispecies Studies and the Microbiome." *Environmental Humanities* 8 (1): 57–76.

———. 2017. "Parasites, Ghosts, and Mutualists: A Relational Geography of Microbes for Global Health." *Transactions of the Institute of British Geographers* 42 (4): 544–58.

Mintz, Sidney. 2011. "The Absent Third: The Place of Fermentation in a Thinkable World Food System." In *Cured, Fermented and Smoked Foods: Proceedings of the Oxford Symposium on Food and Cookery*, edited by Helen Saberi, 13–29. Blackawton, England: Prospect Books.

Paxson, Heather. 2008. "Post-Pasteurian Cultures: The Microbiopolitics of Raw-Milk Cheese in the United States." *Cultural Anthropology* 23 (1): 15–47.

Phillips, Lynne. 2006. "Food and Globalization." *Annual Review of Anthropology* 35 (September): 37–57.

Pollan, Michael. 2013. *Cooked: A Natural History of Transformation*. New York: Penguin Books.

Rabinow, Paul, and Nikolas Rose. 2006. "Biopower Today." *BioSocieties* 1 (2): 195–217.

Rueda Esquibel, Catriona, and Luz Calvo. 2013. "Decolonize Your Diet: A Manifesto." *NSN: nineteen sixty-nine; an ethnic studies journal* 2 (1): 1–5.

Sarmiento, Eric. 2015a. "Umwelt, Food, and the Limits of Control." *Emotion, Space, and Society* 14 (February): 74–83.

———. 2015b. "The Local Food Movement and Urban Redevelopment in Oklahoma City: Territory, Power, and Possibility." PhD diss., Rutgers University. ProQuest/UMI.

Schneider, Gregory W., and Russell Winslow. 2014. "Parts and Wholes: The Human Microbiome, Ecological Ontology, and the Challenges of Community." *Perspectives in Biology and Medicine* 57 (2): 208–23.

Slashinski, Melody, Sheryl McCurdy, Laura Achenbaum, Simon Whitney, and Amy McGuire. 2012. "'Snake-Oil,' 'Quack Medicine,' and 'Industrially Cultured Organisms': Biovalue and the Commercialization of Human Microbiome Research." *BMC Medical Ethics* 13 (28): 1–8.

Slocum, Rachel. 2007. "Whiteness, Space, and Alternative Food Practice." *Geoforum* 38 (3): 520–33.

Spackman, Christy. 2017. "Formulating Citizenship: The Microbiopolitics of the Malfunctioning Functional Beverage." *BioSocieties* 13 (1): 41–63.

Speake, Stephen. 2011. "Infectious Milk: Issues of Pathogenic Certainty within Ideational Regimes and Their Biopolitical Implications." *Studies in History and Philosophy of Biological and Biomedical Sciences* 42 (4): 530–41.

Troitino, Christina. 2017. "Kombucha 101: Demystifying the Past, Present and Future of the Fermented Tea Drink." *Forbes*, February 1, 2017. https://www.forbes.com/sites/christinatroitino/2017/02/01/kombucha-101-demystifying-the-past-present-and-future-of-the-fermented-tea-drink/#6bda579b4ae2.

Via Campesino. 2009. "Nyéléni Declaration." *Journal of Peasant Studies* 36 (3): 673–76.

Williams, Thomas Chatterton. 2017. "The French Origins of 'You Will Not Replace Us.'" *New Yorker*, December 4, 2017.

Winter, Michael. 2003. "Embeddedness, the New Food Economy, and Defensive Localism." *Journal of Rural Studies* 19 (1): 23–32.

The Spandrels of San Marcos?   13

On the Very Notion of Landscape
Ferment as a Research Paradigm

Vaughn Bryan Baltzly

Fermentation provides this volume's organizing motif. The claim is that the biochemical process of fermentation supplies an apt metaphor for understanding certain kinds of landscape change. The kinds of landscape change in question are fortuitously those widely thought to be frequently occasioned by commercial processes centered on the literal metabolic activity of microorganisms breaking down sugars to produce organic acids, gases, and (most famously) alcohol: the commercial production of beer, wine, spirits, cider, cheese, and related fermented products. But what makes this metaphor apt? In this chapter I offer a number of considerations germane to an evaluation of the "fermented landscapes" research paradigm—considerations to which readers may wish to remain alert as they study the chapters herein and as they digest the overall contribution made by this volume.

Specifically, I offer the following three-question sequence as a framework for assessing the merits of this research program:

1. Might "fermentation" refer to a distinctive kind of landscape change? (Or is it better understood simply as a synonym or metaphor for landscape change in general?)

If the answer is yes, we can proceed to ask this next question:

2. Is there any covariance between the appearance of fermentation-centered industries (those centered on beer, wine, spirits, and the like)

and the advent of fermentation-modeled landscape change? (Or do nonfermentation-centered industries frequently contribute to landscape ferment as well? And do fermentation-centered industries frequently contribute to nonfermentation-modeled landscape change too?)

If the answer is yes, we can then proceed to ask a third question:

3. Is there any particular reason why these sorts of industries might be associated (at rates greater than chance) with fermentation-modeled landscape change? (Or do we have here little more than happy linguistic coincidence?) And if so, is there any essential (or at least interesting and surprising) connection between *literal* (chemical and metabolic) fermentation and *metaphorical* (landscape) fermentation? (Or is this again just a matter of verbal happenstance?)

An affirmative answer to even just the first of these questions would seemingly offer some degree of vindication to the notion of landscape ferment as a research program (although this is not to imply that uniformly negative answers to these questions entail that the paradigm lacks merit!). An affirmative answer to the first question would suggest that the research paradigm of fermented landscapes (FL) is *illuminating*; an affirmative answer to the second question would suggest that FL is *fruitful*; and an affirmative answer on the third question would suggest that FL is *probative*.

We shall explore each of these questions in somewhat more detail in what follows. First, though, it is worth taking a bit of space to develop—at least as a foil—the position of a skeptic who holds that the notion of landscape ferment fails to capture any meaningful phenomenon or to otherwise advance our understanding in any helpful manner. The particular skeptical position I shall develop holds that the FL paradigm is simply an instance of what one might term mere "spandrel scholarship." Appropriating a notion from evolutionary biology (where it was first expressed in Stephen Jay Gould and Richard Lewontin's famous 1979 paper, "The Spandrels of San Marco and the Panglossian Paradigm: A Critique of the Adaptationist Programme"), I classify as an academic spandrel any research paradigm or body of work that emerges as a by-product of that familiar and altogether indispensable feature of human understanding and academic discourse: metaphor. So, before turning to a study of each of the three alternative hypotheses suggested by the three questions above, let us begin by exploring this "null hypothesis."

H_0: FL Is Mere Spandrel Scholarship

In 1979 the biologists Stephen Jay Gould and Richard Lewontin adopted the architectural term "spandrel" to refer to phenotypic traits of organisms that, while not themselves evolutionarily adaptive, nevertheless genetically covaried with other traits that *did* confer reproductive advantage and that therefore *did* survive the process of natural selection. Spandrels in this sense are evolutionary by-products of the process of natural selection. Furthermore, a spandrel retains its status as such even if and when that trait later proves adaptive (or, to use the term Gould and his colleague Elisabeth Vrba [1982] later coined for this very purpose, *exaptive*) for some other purpose.

In architecture the term "spandrel" is the name for the solid surface occupying the space between the shoulders of adjoining arches and the ceiling or molding above, or the space between two arches or between an arch and a rectangular enclosure. Such surfaces (and other related architectural forms, such as pendentives) originated as necessary by-products of essential design features responsive to fundamental architectural constraints—in the case of spandrels, the necessity of employing arches to support a dome. However, they later became architecturally "adaptive" in their own right, as they proved useful in supplying surfaces on which to engrave or paint iconography—typically (since spandrels are a prominent feature in Gothic cathedral architecture in particular) depictions of the saints, scenes from the Gospels, and related Christian imagery. (In a decision that lent their 1979 paper its title, Gould and Lewontin used the example of San Marco Basilica in Venice, the spandrels of which the reader can glimpse in figure 39.)

Gould and Lewontin co-opted the notion of spandrel from the

FIG. 39. The "spandrels" in San Marco (St. Mark's) Basilica in Venice, which inspired the title of Stephen Jay Gould and Richard Lewontin's famous 1979 paper. They described these triangular, dome-supporting structures as spandrels, but technically they are better characterized as pendentives. Photo by Vaughn Bryan Baltzly.

field of architecture in the service of their critique of a then-prominent position in their own field: what they termed the "adaptationist programme." Adaptationists, they suggested, tended to see any conceivable isolable trait of every organism as a fitness-enhancing adaptation. This unrestrained enthusiasm for "adaptationism" led many Anglophone evolutionary theorists to see evidence of evolutionary optimization everywhere—much as we observe the Pollyannaish Pangloss in Voltaire's *Candide* mistakenly inferring that the bridges of our noses were designed to hold up our spectacles.[1] Classic examples of evolutionary spandrels include "masculinized genitalia in female hyenas, exaptive use of an umbilicus as a brooding chamber by snails, the shoulder hump of the giant Irish deer, and several key features of human mentality" (Gould 1997, 10750). These "key features of human mentality," often cited in this regard by Gould and many other thinkers, include music and perhaps even language itself (see, e.g., Buss et al. 1998).

So language itself may be an evolutionary spandrel. (Although for a dissenting perspective on the alleged "spandrel-ity" of language, readers might consult some of the work of Stephen Pinker, e.g., Pinker and Bloom 1990; Pinker and Jackendoff 2005.) Irrespective of the evolutionary status of language—as either adaptation, exaptation, or (mere) spandrel—it nevertheless seems clear that language itself generates its own sort of spandrels. Metaphor provides one striking site of this possibility. Undoubtedly the use of metaphor often helps to advance our understanding by, for example, illuminating some facet of an issue or phenomenon or by suggesting parallels between phenomena that otherwise appear to be quite disparate, and so on. But otherwise-illuminating metaphors often carry with them a surfeit of meaning, thereby offering enterprising authors ample opportunity to indulge their artistic abilities in exploiting to the fullest degree the literary possibilities contained within the metaphor. Like the architectural support structures of Gothic cathedrals—once upon a time aesthetically inert but later recognized by enterprising artists as providing great artistic potential—certain metaphors offer to certain scholars a similar potential. In other words, just as medieval artists seized the opportunity to embellish the functionally and structurally necessary spandrels—exploiting a new platform for showcasing their craftsmanship—so also do some contemporary scholars seize the opportunity to exploit the full literary potential of what (in and of themselves) are useful metaphors.[2] To demonstrate their technical and conceptual mastery, artists and scholars alike have appropriated devices that do indispensable "structural" work—whether with respect to the physical architecture of an edifice or to the conceptual architecture of an argument or research program—and have mined them for all their artistic and literary worth.

It is important to note a relevant trichotomy that presents itself at this point. In evolutionary thought, a given trait may be regarded as, first, what I shall term here "mere spandrel" (a by-product of evolution: nonadaptive in its own right but naturally selected nevertheless, owing to its genetic association with an originally adaptive trait); second, an exaptive spandrel (a spandrel that is later drafted into some other use that does enhance reproductive fitness); or third, originally adaptive (that is, as having arisen as a stable feature of a population due to that trait's having conferred genuinely adaptive advantage upon organisms possessing it). Thus, to label a trait a spandrel is not necessarily to condemn it for its disutility; it may well have proven exaptive. Matters stand likewise with respect to the notion of spandrels as applied to academic research programs. A metaphor-based paradigm or analytic framework may be a "mere spandrel" (a by-product of other [admittedly useful] research paradigms, which nevertheless fails to advance our understanding or knowledge in its own right). Such a framework may be exaptive (one that perhaps initially arose as a by-product of some other research paradigm but that later becomes autonomous and proves useful in its own right).[3] Finally, it might be originally "adaptive" (that is, one that initially arose because it could produce improved understanding and insight in its own right). The same moral should therefore be drawn with respect to the "spandrelity" of research programs or of individual acts of scholarship: to demonstrate that a paradigm or paper originated as a sort of by-product of a metaphor's surplus meaning is manifestly *not* to judge it incapable of effecting genuine advances in our knowledge or understanding; it may in fact prove "cognitively exaptive." To think otherwise is to commit the genetic fallacy, or the fallacy of origins: to draw (in the formulation of that fount of collective knowledge known as Wikipedia) a "conclusion that is based solely on someone's or something's history, origin, or source rather than its current meaning or context."[4] For now we need not be concerned with the distinction between adaptive and exaptive research paradigms (whatever precisely such distinction might amount to); we need only investigate that which distinguishes "mere spandrel" programs from those that are genuinely illuminating, fruitful, or probative.[5]

Some readers may protest that I've gone too far. They may allege that the present invocation of "spandrel scholarship" is itself an instance of spandrel scholarship. By unpacking the full conceptual and literary potential latent in the notion of a spandrel—by excavating various layers of meaning first conferred upon the notion when the contemporary evolutionary biologists appropriated the medieval cathedral architects' term—am I not exemplifying precisely that which I seek to characterize? Perhaps so—in which case I reply, "All the better!" Our judgment as to the notion's utility may be further aided by our assessment of its value in the present case.

But how to make this judgment? It is to that question we now turn. In the remainder of this chapter I offer a number of "alternative hypotheses" that readers might consider when judging whether (or to what extent) a metaphor-based research paradigm is—to continue with *my* organizing motif—more than mere spandrel.

H₁: FL Provides Illumination

One way in which we might fail to reject our null hypothesis is if the ferment concept serves as nothing more than an image of landscape change tout court—if "fermentation" is little more than a synonym for "landscape change." So for now let us adopt as our first research question the first query posed in this chapter's second paragraph: "Might 'fermentation' refer to a distinctive kind of landscape change?"

If we judge the answer here to be no, then we are committed to regarding FL as mere academic spandrel. (Again: it is worth repeating that this is not necessarily to condemn it as lacking all utility. Perhaps there is considerable literary talent on display in this volume's various efforts to play with this notion; maybe it has artistic value as a certain kind of academic or conceptual poetry.) If on the other hand we decide that the answer to this question is yes, then we have judged FL to be illuminating. That is, the metaphor that stands at its center may itself represent a taxonomic advance. If there are species of landscape change usefully conceptualized in terms of ferment, then it may well turn out that introducing this term into the academic literature draws scholars' attention to a heretofore undiscovered (or at least underappreciated) subset of landscape change. So how might we proceed in evaluating our first research question?

One way to test whether "fermentation" captures a distinctive subset of landscape change is to ask whether we can point to some form of (physical and/or sociocultural) landscape change that cannot be meaningfully or helpfully conceived in the image of fermentation. If not, then we're not being directed to any particular *kind* of landscape change; we're just learning that fermentation-centered industries often play a role in bringing about landscape change. In this regard it is of course important to ask what other kinds of landscape change there even are. As a nonspecialist in this regard, I can only propose a range of naïve alternatives drawn mainly from the vernacular of popular commentary: gentrification, industrialization, decay, restoration, urbanization, residential development, and the like. If these are even the right sorts of things to count as alternative models of landscape change, then it seems that landscape ferment might be a distinctive species of the wider genus.

Next we need to ask what in particular distinguishes those episodes of landscape change modeled on ferment from these other varieties. One thought springs naturally to mind here. Suppose we take our metaphor seriously. In other words, suppose we focus on the imagery suggested by having literal, metabolic ferment (as opposed to fermentation in its broader, more generic sense of "agitation [or] excitement") serve as a model for landscape change.[6] If we do, we're apt to focus on the notion of "microagents" effecting a multitude of small-scale, "local" changes, the cumulative effect of which is to transform the local environment into something new (and perhaps into something better and more exciting?). This might notably differ from, for example, large-scale regional planning of the top-down variety—as we might observe with a residential subdivision. Landscape ferment may also differ in systematic ways from ordinary suburban sprawl, insofar as the former (but not the latter) involves microagents figuratively "consuming" the available "raw materials" of the local landscape, transforming those materials (and thereby that landscape), and as a result "secreting" a genuinely new, unified, and highly desirable by-product, like wine country or an urban beer trail. Whereas with ordinary suburban sprawl, what results is not unified or "consumable"—as are, say, wine country viewsheds and other "consumable landscape" tourist destinations—but rather diffuse and undifferentiated. (Chapters 2 and 3 of the present volume include some case studies illustrating how various geographic regions have successfully [or, in the case of the Canterbury Plains region of New Zealand, unsuccessfully] manifested such bottom-up place-making dynamics.)

Thus, if the phrase is helpful in capturing a distinctive subset of landscape change, then the metaphor (and the associated research paradigm) may prove illuminating. But this may seem especially so if the advent of fermentation-based industry tends to be linked to the phenomenon of fermentation-modeled rural and/or urban development—that is, if there appears to be covariance between the phenomena. Thus it is to this next alternative hypothesis—that suggested by the second question from the second paragraph of the chapter—that we now turn.

H_2: FL Is Fruitful

Another way in which we might reject our null hypothesis is if the phrase "fermented landscapes" points to some robust relationship between fermentation-centered industries and fermentation-modeled landscape change. So let us now adopt as our second research question the second query posed above: "Is there any covariance between the appearance of fermentation-centered industries (those centered on beer, wine, spirits, and the like) and the advent of fermentation-modeled landscape change? (Or do nonfermentation-centered industries frequently contribute to landscape ferment as well? And do fermentation-centered industries frequently contribute to nonfermentation-modeled landscape change too?)"

If we answer this question in the negative, we've decided that FL—while illuminating—fails to be fruitful. In other words, in classifying types of landscape change, fermentation captures a distinct category—it just turns out that this category of landscape change isn't only or always (or even often) associated with industries of ferment. If we answer this question in the affirmative, though, we've seemingly identified a fruitful avenue for further exploration and investigation; we will have discovered (to anticipate our subsequent discussion) reason to proceed to our third research question, as well as to our third alternative hypothesis: theorizing as to why such a connection obtains.

Ascertaining the strength of the covariance between the proliferation of commercial enterprises centered on fermented goods (what I shall term "commercial ferment" for short) and landscape ferment is principally an empirical matter, requiring careful definition, diligent observation and data collection, and statistical acumen. Accordingly, it likely falls outside the scope of the present volume and its individual contributions, the purpose of which is primarily to lay the groundwork and indicate some promising directions for future research.[7] Enthusiastic empirical investigators studying this volume might, however, find its contents sufficiently illuminating and suggestive that they may be inspired to formulate research projects and secure grant money to explore further the full extent of the hypothesized connection. Until such time as those studies are completed, we might have to simply offer a promissory note: "further investigation is required" before we can determine whether there is a robust statistical correlation between the two.

But the previous paragraph also points toward another sense of "fruitfulness" relevant here: the FL paradigm may serve to unify seemingly disparate strands of scholarship, revealing otherwise unrelated research programs to be connected in surprising and fruitful ways. The discovery of such connections too may also prove generative of further research, as might the creative prompt of the very notion of landscape ferment itself. In many ways we might hope to evaluate the paradigm's fruitfulness via the merits of the present volume's contents. And indeed the very diversity and wide-ranging-ness of the contributions assembled within these pages seems indicative—at least as a first approximation—of FL's fecundity.

It may be worth noting that there are (at least) two different ways in which researchers may conclude that there is a less than fully robust covariance between episodes of commercial ferment and landscape ferment. The first is if they uncover a paucity of co-occurrence. Suppose it turns out that, once the notion of landscape ferment becomes suitably operationalized so that it can be studied and measured, researchers discover that instances of landscape fermentation only rarely accompany localized outbreaks of fermentation-centered commercial activity. In that case we may have no choice but to demur at the suggestion of the paradigm's fruitfulness on account of this observed paucity. Alternatively, researchers may soon discover an abundance rather than a dearth of such co-occurrences, though only because there is an abundance of landscape fermentation in general. Suppose it turns out that landscape ferment frequently arises in response to a wide and diverse array of commercial enterprises—as well as perhaps in response to other sorts of noncommercial stimuli. If the causal antecedents of landscape ferment prove too heterogeneous, the notion may lose much of its fruitfulness as a lens for studying fermentation-centered, industry-driven landscape change (though it may not lose all of its fruitfulness in this regard: presumably it would still be an interesting question as to why fermentation-centered industries are disproportionately associated with fermentation-modeled landscape change, rather than, for example suburban sprawl).

In any event, supposing for the moment that we concur as to FL's fecundity, let us proceed to our third alternative hypothesis—that suggested by the third question from the chapter's second paragraph—regarding the paradigm's explanatory power.

H₃: FL Has Explanatory Power

One further way in which we might reject our null hypothesis is if the notion of ferment plays any useful role in explaining the (possibly) observed covariance between fermentation-centered commercial activity and fermentation-modeled landscape change (as investigated pursuant to H_2 above)—if it serves to unify these two usages (one literal, one metaphorical) of the term "ferment." So let us now adopt as our third research question the third query posed above: "Is there any particular reason why these sorts of industries might be associated (at rates greater than chance) with fermentation-modeled landscape change?" If we answer this question in the negative, we have decided that FL— while fruitful—is not probative: that what we have here is little more than happy linguistic coincidence. If we answer this question in the affirmative, however, we have judged that FL has actual explanatory power, insofar as it uncovers underlying mechanisms or relations linking these two forms of fermentation—or at the very least insofar as it renders sensible this observed relationship.

What is the strongest case one might make for the probative power of this metaphor? Here we might further develop the microagents analogy first suggested in our discussion of H_1 above: like the chemical process of fermentation, certain forms of landscape change occur as the aggregate of lots of (uncoordinated) microchanges, performed by independent microagents, not operating under the auspices or direction of any one overarching or guiding (macro-)agent of change. Such evolutions in the local landscape are more likely to be driven by actors who understand themselves according to the models of small-scale, local, and/or microproducers. These days, at least, fermentation-centered enterprises disproportionately conceive of themselves in such fashion. (Consider the current enthusiasm for microbrews, small-batch brewing, craft beers [and wines and spirits] as pushback against the perceived conformity and homogeneity of Big Beer and "macrobrews," etc.[8]) Given these trends in the beer, wine, and spirits industry, and in the wider current culture (pun semi-intended), it would come as little surprise if fermentation-based enterprises should turn out to be at the heart of many prominent instances of fermentation-esque landscape change.

Of course, this raises a further question: Is there any particular reason why the current mania for "local," "craft," and "micro" should obtain in the market for booze to an extent greater than it does in other markets? Is there any essential connection between the role of *fermentation* in these industries and the aforementioned craze (which gives rise to the advent of so many microactors in this space)? The answer is almost certainly no—this is a contingent connection (as evidenced by the fact that similar crazes characterize other *non*-fermentation-centered markets—such as that for food more generally).[9]

So, given what we already understand about current consumer trends surrounding fermented products, it should perhaps be unsurprising that we would observe the linkage at the center of the FL paradigm. But is this anything more than happy linguistic coincidence? Here perhaps we may have finally found the limits of our metaphor. I doubt very much that there is any essential (or interesting, or surprising) connection between literal (metabolic) fermentation and metaphorical (landscape) fermentation. As already noted, there is no particular reason why we might expect a "microagent pedigree" to be any more or less fashionable among consumers in the fermentation space than it is anywhere else. This appears to be simply a contingent connection characteristic of our current culture; only time will tell if it proves to be a stable, more or less permanent feature of consumer demand for fermented products or just a momentary fad. Time will tell, that is, whether this fashion (like so many others) will soon recede or whether it will (as fashions so often do) spread to other segments of the commercial marketplace as well—such that someday we may observe landscape ferment to be widely effected by increased commercial activity in the (small-batch) garment industry, or in the (micro) pharmaceutical industry, or in the (artisanal) publishing industry, or in the (craft) bicycle industry, and so forth.

Conclusion

Having raised three questions for the reader's consideration, I shall now hazard my own tentative answers—though, I hasten to add, not in any particular effort to persuade readers to share my views. With respect to the three questions articulated in this chapter's second paragraph, I suspect that we might answer as follows:

1. **Yes.** It seems quite plausible that we can and should contrast the concept of landscape ferment with, for example, industrialization, residential development, urban sprawl, and the like. Accordingly, one interesting question for geographers and urban planners to ponder in this regard is the following: Is gentrification best understood as a species of (urban) landscape ferment?

2. **I do not know.** Answering this question lies outside my academic expertise; professional geographers will have to guide me here. However, if they convince me that there is in fact a correlation between episodes of commercial and landscape ferment, I think I'd know how to answer the next question.

3. **Yes**, with respect to the first part of the question. Consider the microagents analogy: biochemical fermentation results from the aggregate effect of the uncoordinated consumptive behaviors of large numbers of microagents. Matters stand likewise with landscape ferment, which typically transpires unguided and unenvisioned by any overarching macroagent. The analogs of yeast and bacteria in this case are the commercial enterprises self-consciously styled as small, family, craft, and local. But with respect to the latter part of the question, the answer seems to be no. It strikes me at this point that we have here merely a verbal parallel. As we just observed, the taste or fashion for small/craft/family/local varieties of a product can arise with respect to practically any industry; there's nothing unique about fermentation-centered enterprises in this regard.

We have arrived at last at the skeptical challenge implied (if not outright articulated) in our titular question: Are the contributions to this volume equivalent to the (academic) spandrels of San Marcos (Texas)? (San Marcos is the home of Texas State University—a fine institution [says this biased observer] whose scholars are heavily represented within these pages.) Is the "fermented landscape" paradigm merely the by-product of a metaphor—simply a canvas upon which enterprising scholars of food, drink, and/or landscape change can exhibit their conceptual dexterity?

This reader answers with a fairly confident "no": the contributions to this volume do constitute a legitimate and coherent research paradigm. The fact that it arises from the exploitation of a fortunate verbal parallel (between the applications of two ordinary senses of the English word *ferment*) does nothing to detract from the paradigm's illuminating, fruitful, and probative properties. To press the analogy with the evolutionary biologist's sense of "spandrel" a bit too far, perhaps, we might say that landscape ferment is neither (mere) spandrel nor (merely) exaptive; it is instead fully adaptive in its own right. Whatever the precise etiology of the particular linguistic "mutation" at the heart of this paradigm— someone somewhere coined the bon mots "fermented landscapes" and a research paradigm was born[10]—the paradigm born of this mutation has immediately proven to be adaptive in its own right, as I believe the contents of the present volume amply demonstrate.[11]

Ultimately, however, it is not *this* reader's judgment that concerns me most. The titular question is one readers must answer for themselves— *you*, and all your colleagues and students who are also studying this volume. I simply hope to have provided a helpful framework for thinking these matters through.

Notes

For formative conversations on the very notion of "spandrel scholarship," I am grateful to Bob Fischer and Anthony Cross, both of Texas State University's philosophy department. (I am not sure, however, whether either Bob or Anthony would endorse the use to which I've subsequently deployed the notion!)

1. Gould and Lewontin (1979) believed the error to be less pervasive among continental biologists.

2. Andy Murray's contribution to this volume (see chapter 11) might even be read as a diagnosis of the way in which certain parties currently exploit the term "fermentation" in something like this fashion.

3. It is tempting to view the birth of computer science in something like this light. Late nineteenth-century and early twentieth-century advances in formal logic generated what arguably were spandrel-like metalogical research programs dedicated to exploring the properties of, for example, decidability, provability, and completeness as they applied to first-order formal systems. In the hands of thinkers like Kurt Gödel, Alonzo Church, and Alan Turing, these programs morphed into something that (especially when merged with concurrent advances in electronic circuitry) proved exaptive as the mathematical and logical foundations of computer science. If this view of matters is even tolerably accurate, just think how much the contemporary world has been shaped by academic spandrels!

4. "Genetic fallacy," Wikipedia, accessed November 7, 2018, https://en.wikipedia.org /wiki/Genetic_fallacy.

5. Or, for those familiar with the term used by Daniel Dennett (2006), we are interested in criteria distinguishing illuminating, fruitful, and probative research paradigms from those that are akin to investigations of chmess, the chess-like game in which the king may be moved two spaces instead of one.

6. It's worth noting that what we take to be the literal sense of "metabolic ferment" is, unsurprisingly, historically contingent. For a nuanced discussion of how our current understanding of fermentation is in fact historically conditioned in all sorts of interesting ways, see Andy Murray's contribution (chapter 11) to the present volume. "Agitation [or] excitement" comes from the third definition of fermentation at dictionary.com (the online dictionary first elucidates the organic chemist's current sense of the term), as of November 26, 2018.

7. However, see chapter 4 (by Holtkamp, Lavy, and Weaver) in the present volume for interesting and substantive steps in this direction.

8. The nature and extent of this enthusiasm are further analyzed and documented in chapter 2 of this volume.

9. An interesting further question would be why we observe this feature of consumer demand so prominently in the market for food (including of course drink) but not in other places. Why is there no comparable mania for locally sourced, small-batch, or micro versions of such things as medicine, furniture, clothing, or consumer electronics (see Myles and Baltzly 2018)?

10. In the present case we do of course know the identity of this "someone somewhere": it was Colleen Myles, in 2016. So although the source is known in this case, some-

times the originators of metaphor-driven research paradigms are—like the genetic mutations that prove biologically adaptive or exaptive—lost to the mists of time.

11. Readers interested in the wider application of evolutionary concepts as metaphors or analogies for understanding the fates and fortunes of research paradigms are invited to consult the seminal discussion of memes (in the original meaning of that term) in the work of Richard Dawkins (1976, chap. 11). Dawkins offers an illuminating discussion of the ways in which ideas (and phrases, and fashions, and paradigms, and other cultural artifacts) can themselves be understood in parallel with genes—as "units of selection" whose survival and proliferation are proportional to their cultural or intellectual adaptiveness. According to Wikipedia's entry on memes, this notion that ideas and fashions are subject to the same selective pressures as genes is not original with Dawkins but was discussed even during Darwin's day. T. H. Huxley (1880, 1) is quoted there as writing that "the struggle for existence holds as much in the intellectual as in the physical world. A theory is a species of thinking, and its right to exist is coextensive with its power of resisting extinction by its rivals."

References

Buss, David M., Martie G. Haselton, Todd K. Shackelford, April L. Bleske, and Jerome C. Wakefield. 1998. "Adaptations, Exaptations, and Spandrels." *American Psychologist* 53 (5): 533–48.

Dawkins, Richard. 1976. *The Selfish Gene*. New York: Oxford University Press.

Dennett, Daniel C. 2006. "Higher-Order Truths about Chmess." *Topoi* 25 (1–2): 39–41.

Gould, Stephen Jay. 1997. "The Exaptive Excellence of Spandrels as a Term and Prototype." *Proceedings of the National Academy of Sciences* 94 (20): 10750–55.

Gould, Stephen Jay, and Richard Lewontin. 1979. "The Spandrels of San Marco: A Critique of the Adaptationist Programme." *Proceedings of the Royal Society of London: Series B, Biological Sciences* 205 (1151): 581–98.

Gould, Stephen Jay, and Elisabeth S. Vrba. 1982. "Exaptation: A Missing Term in the Science of Form." *Paleobiology* 8 (1): 4–15.

Huxley, T. H. 1880. "The Coming of Age of the Origin of Species." *Nature* 22:1–4. https://www.nature.com/articles/022001a0.

Myles, Colleen C., and Vaughn Baltzly. 2018. "Why *Food*, Though?" Presentation at Exploring Ethics through Food Choices: Fourth Biennial Perugia Food and Sustainability Studies Conference, Umbra Institute, Perugia, Italy, June 8, 2018.

Pinker, Stephen, and Paul Bloom. 1990. "Natural Language and Natural Selection." *Behavioral and Brain Sciences* 13 (4): 707–27.

Pinker, Stephen, and Ray Jackendoff. 2005. "The Faculty of Language: What's Special about It?" *Cognition* 95 (2): 201–36.

On the Future of Fermented Landscapes as a Focus of Study

14

Colleen C. Myles, Walter W. Furness, and Shadi Maleki

Macro Consequences for Micro(be) Processes

Fermentation is a concept—and a process—that is laden with meaning, both metaphorically and materially. Throughout this text we have demonstrated how the biochemical process of fermentation fits the trialectic of *meaning-model-metaphor* (Hiner 2016) in that it has meaning as a form of metabolic transformation, it serves as a model for the kinds of landscape change and development we observe as related to fermentation industries, and it provides an apt metaphor for broader processes, providing additional insights beyond simple description. In this volume we have highlighted the significance of fermentation in these various forms to medicine, food and drink, biotechnology, nutrition, landscape change, and sociocultural production and have suggested a research agenda that contextualizes and questions the spaces and places of fermentation at both macro and micro scales (chapter 1). The concept of fermented landscapes is meant to make room for creative reflection and analysis of the varied developments we see in rural, urban, and somewhere-in-between places—developments and changes that have led to both expected and unexpected benefits and challenges for people and landscapes all over the world.

What can these different perspectives and snapshots offer us? Perhaps, as Michael Pollan (2013, 297) suggests (via Lynn Margulis), they may open our eyes to the "unseen universe of microbes all around and within us" while enlivening and making visible the innate position(s) of fermentation in the human experience (chapters 10 and 12). From this point of departure the lens of fermented landscapes brings a holistic approach to sometimes overlooked relationships and processes. In short, the mundane is not banal; everyday activities and products related to fermentation are replete with (contested) meaning and nuance, and humans are far from the only actors at work (chapters 10 and 11).

In reflecting upon the cases and claims presented here, we return to the metaphor of fermentation vis-à-vis landscape change and ponder how this notion may be an apt one (chapter 13). First, is there a distinctive form of change at play beyond mere euphemism? Second, are there significant linkages between the production of ferments and the production of fermented landscapes? Third, if so, why? What are the connections that transcend convenient semantic coincidence? To explore the concept's veracity, in this volume authors have asked questions like the following:

- What does it mean to consider fermented landscapes as an organizing theme for research and analysis? Are there material and metaphorical dimensions of change that might be (better) understood vis-à-vis the concept of fermentation? (chapter 1)
- Although fermentation-focused regions are almost universally celebrated or pursued, not all are successful—nor are all universally beneficial. Thus:
  - ▷ What elements of fermentation-focused development, for better or worse, are "private" and which are "public" when the boundary lines between what is (non)excludable and (non)rivalrous are sometimes blurrier than they appear? (chapter 2)
  - ▷ And, more specifically, why is it that some aspiring wine (or beer or spirit . . .) regions thrive while others shrivel? (chapter 3)
- What environmental, cultural, or political histories or aspirations drive "booze booms" of various sorts? (chapters 4, 5, 6, and 7)
- What other (nonalcoholic) ferments can (help) explain landscape change and the communities within which it occurs? (chapters 8 and 9)

- For the more "micro" processes at work, what are the more figurative elements of fermentation that can guide our thinking about small-scale universes, research, and knowledge production more broadly? (chapters 10, 11 and 12)
- And, finally, given that the concept of fermented landscapes seemingly has some analytical purchase (as evidenced by the contents of this volume), is there some particular explanatory power offered by fermentation as a lens for viewing and understanding landscape change and/or development? (chapter 13)

Summary of Findings

In the first section of the book the opening chapter outlines the contours of the fermented landscapes framework. The second discusses the collective "goods" and externalities of fermentation-focused development. In chapter 3 John Overton describes how narratives of triumph and success define the geography of wine literature, but he argues that "risk, struggle, debt, setbacks, and failure are as much a part of the narratives of winemaking as are awards, quality, popularity, and success." Overton contends that the fermentation metaphor in "fermented landscapes" is apt because while there are necessary factors for a (budding) wine region to succeed, those factors are not sufficient; time and an element of random chance, whereby luck or misfortune, fortuitous or bad timing "can tilt outcomes one way or the other," are also key to the process. He supports the claims made in chapter 2 about the "public good" elements of place-making and fermentation-focused development by outlining precisely how important such collective imaginaries are in establishing (regional) success. Overton concludes by noting that "'fermented landscapes' are not just reflections of the products they produce; they are themselves the result of a complex alchemy."

In chapters 4 through 9, which encompass the second section of the book, the authors examine a number of landscapes of (different) ferment(s). Christopher Holtkamp et al. discuss bourbon in Kentucky and beyond (chapter 4), noting the spread of the product and its associated identities as related to the physical environmental factors necessary for its production. Chapter 5 highlights cider in England, especially as related to varying, articulated scales of operation, surprising forms of cooperation and competition, and the interconnections among actors of all kinds. Next the authors explore beer geographies, or "beerscapes," in the United States as seen via various cultural migrations (chapter 6). Chapter 7 follows the evolution of Bloody Mary cocktails in the United States as a reflection of the "taste of place" in various locales, while Ryan Galt examines farm-to-bar and bean-to-bar chocolate in Hawaii (chapter 8), with special attention paid to the role of quality and agritourism in the industry's success. Finally, Elizabeth Yarbrough, with Colleen Myles and Colton Coiner, explores the role of kombucha in re-enlivening culture (of various kinds) and communities in San Marcos, Texas (chapter 9).

In the final section of the volume, the authors delve deeper into the microprocesses of material and metaphorical fermentation. Maya Hey (chapter 10) explores the meaning(s) of similar things, like milk or bacteria, in different spaces (e.g., in kitchens or laboratories or on hands) to interrogate their "complex materiality" and to locate agency in human and more-than-human loci. She writes, "Fermentation is a process of transformation due to microbes like bacteria, molds, and yeasts that transform food ingredients both materially and symbolically; the resulting ferment is just as much a product of time and ambient factors as it is a product of human and microbe interactions." Offering a view similar to Overton's conclusion at the landscape scale (chapter 3), Hey recognizes random chance as much as human (or more-than-human) agency as part of the equation for ferment at the micro level. This realization gives pause to notions of control in emerging biotechnical approaches to fermentation.

Andy Murray (chapter 11) describes conceptions of fermentation and draws comparisons to "craft" fermentation projects from bioengineering fields or practices. He describes how these industrial uses are being framed as fermentation due to the popularity of craft ferments and their perception as "safe," long-standing, and overall culturally and morally acceptable human endeavors. He counterposes these "traditional" ferments with more controversial genetic modification exercises that use some foundationally similar processes or components. Murray ultimately warns of an overstretching of the "fermented" metaphor, which might obscure more meaning than it uncovers.

Eric Sarmiento (chapter 12) similarly argues that not all ferments are equivalent, noting that in today's commercialized, global market some ferments are now pasteurized, stripping them of their "lively" qualities. He discusses how microbial actors are being exploited, a topic Murray (chapter 11) also explored, and argues for a political ecology of fermented landscapes, especially for investigations of our "extensive, intensive ecological tapestries" that are more biopolitical and/or microbiopolitical. He notes that, in a sense, studies of fermented landscapes, while perhaps seen as mainly utilitarian for the macro (landscape) scale, can also be useful for studying microspaces and/or environmental degradation within.

Finally, in the penultimate chapter, Vaughn Bryan Baltzly (chapter 13) asks how and why we might consider a fermented landscapes metaphor. He works through how readers might conduct a stepwise evaluation of the idea to discover how and why the concept might be illuminating, fruitful, and probative or explanatory. He concludes that the answers to these inquiries are mainly affirmative.

Fermented Landscapes Research, Prospectively

Fermented landscapes research seeks to discover the possibilities of place creation, asking: How does place creation via fermentation produce a public good? How uneven is this good? (chapter 2). What are the implications for economic and social justice? What about neighborhood/community change and gentrification? How does the life cycle of craft projects effect positive and negative externalities in the landscape? One needs to look no farther than these pages to see that some fruitful studies have already been conducted along these lines of inquiry. However, there is more work to be done. Several ideas are currently germinating within this research group, including interest in a broader study of craft industry (e.g., beer) that deconstructs each component (water, yeast, hops, grain, adjuncts) by tracing its sources and social-political dimensions.

As noted within these pages, the world of artisanal fermentation (e.g., craft beer, wine, cheese, etc.) has drawn significant attention from scholars. However, much of this research focuses on questions of place-making, (neo)localism, tourism, economic development, and identity formation. Yet in this growing field some of the smallest actors remain relatively unquestioned. For instance, while certain key ingredients in beer—yeast and hops, for example—constitute only a small percentage of the material weight of the finished product, they are key components of its distinctive character. These tiny contributors carry a heavy percep-tual weight as the creators of sought-after flavors and aromas associated with certain beer styles. Future research by this team will endeavor to better understand the role of these oft-neglected constituents, beginning with an exploration of the place of yeast—specifically, *Saccharomyces cerevisiae* and its wild cousins—in the geography of beer and especially its role as a more-than-human agent in the collaborative process of (co) producing fermented beverages.

Viewing yeast as an active contributor to brewing opens up new ways of understanding human-environment relationships, including questions of world-building associated with terroir (Szymanski 2018). Yeast is central to cultural staples like bread and alcohol (White and Zainasheff 2010), and how we theorize and understand it has cultural, political, and environmental ramifications. With the rise of genetically modified organisms and synthetic biology's Sc2.0 Project, which focuses on a synthesized eukaryotic genome (Richardson et al. 2017; Synthetic Yeast n.d.), questions emerge that are related to how brewers and consumers will respond to the potential availability of yeasts capable of producing hop-like esters and how this may affect the identities and cultural or place-based associations of craft beer writ large, as well as the (formerly?) agricultural production of fermentation inputs (Herkewitz 2014; Szymanski 2017). In short, the microbial life of our food and drink is still being explored, but the significance of these tiny actors is increasingly clear (Paxson 2008; Pollan 2013; chapters 9–12, this volume).

In Sum

Throughout this volume we have continued to bump into instances of collaboration, coproduction, and intra-action between humans and microbes. As noted in several chapters, especially in the last section of the book, a conceptualization (or perhaps recognition) of fermentation as ontologically enmeshed distributes agency across a cast of actors and environmental factors, complicating our notions of independence and power. Instead of neat categories of entities interacting through fermentation, we are challenged to see the nested nature of metabolic relationships and the "liveliness" of nonhuman agents. And yet we are cautioned against applying the fermentation metaphor too liberally, both because it can obscure relevant differences in process or production (chapter 11) and because it may reduce clarity rather than improve it (chapter 13).

With an understanding that the contents of this book simply offer a taste of the work to be done under the auspices of this conceptualization and considering the smattering of ideas we have offered for moving forward, research under the "fermented landscapes" umbrella seems to be ripening for the harvest. This book is intended to be an opening into a conversation that will hopefully continue beyond the perspectives offered herein. In sum, this book covers a lot of ground and offers a compelling glimpse into the future of fermented landscapes rather than providing a definitive, analytical end point. The authors included here—and surely others beyond this cohort—will undoubtedly continue to explore the next frontiers for this research paradigm.

References

Herkewitz, William. 2014. "Scientists Create Synthetic Yeast Chromosome (and Unlock the Future of Beer)." *Popular Mechanics*, March 27, 2014. https://www.popularmechanics.com/science/health/a10289/scientists-create-synthetic-yeasts-and-open-the-door-to-the-future-of-beer-16637455/.

Hiner, Colleen. 2016. "Beyond the Edge and in Between: (Re)conceptualizing the Rural-Urban Interface as Meaning-Model-Metaphor." *Professional Geographer* 68 (4): 520–31.

Paxson, Heather. 2008. "Post-Pasteurian Cultures: The Microbiopolitics of Raw-Milk Cheese in the United States." *Cultural Anthropology* 23 (1): 15–47.

Pollan, Michael. 2013. *Cooked: A Natural History of Transformation*. New York: Penguin Press.

Richardson, Sarah M., et al. 2017. "Design of a Synthetic Yeast Genome." *Science* 355 (6329): 1040–1044. https://doi.org/10.1126/science.aaf4557.

Synthetic Yeast. n.d. "Synthetic Yeast 2.0: Building the World's First Synthetic Eukaryotic Genome Together." Accessed October 14, 2018. http://syntheticyeast.org/.

Szymanski, Erika. 2017. "The Cultivated, the Wild and Everything in Between: Yeast Population Dynamics in the Vineyard and the Winery." *Wine and Viticulture Journal* 32 (6): 12–16.

———. 2018. "What Is the Terroir of Synthetic Yeast?" *Environmental Humanities* 10 (1): 40–62.

White, Chris, and Jamil Zainasheff. 2010. *Yeast: The Practical Guide to Beer Fermentation*. Boulder CO: Brewers Publications.

Contributors

VAUGHN BRYAN BALTZLY is an assistant professor of philosophy at Texas State University, working in political philosophy and other areas of applied ethics. After completing graduate studies in philosophy and in public policy but before coming to Texas State, he spent more than six years employed by the U.S. federal government, working in both the executive and legislative branches.

SAM BATZLI holds a PhD in historical and cultural geography from the University of Illinois at Urbana-Champaign. His interest in data visualization led him to the Space Science and Engineering Center at the University of Wisconsin-Madison, where he manages web mapping and mobile app development for satellite imagery. As a homebrewer he maintains an active interest in the brewing industry and its geography.

COLTON COINER is a Texas State University alum who is interested in place-making.

WALTER W. FURNESS is a PhD student in the Department of Geography at Texas State University. He holds an MS in geography from Northern Illinois University and a BA in environmental science and geography from Northwestern University. His research focuses on political ecology and food geographies, including fermentation, justice, and urban agriculture.

RYAN E. GALT is a geographer and professor in the Department of Human Ecology at the University of California, Davis, and studies agriculture and food systems, with an emphasis on governance and transitions to sustainability. He has examined pesticide-dependent smallholder agriculture in Costa Rica, community-supported agriculture in California, and ethical cacao-chocolate commodity chains.

MAYA HEY is a Vanier Scholar and doctoral candidate in communication studies at Concordia University in Montreal. Her dissertation research focuses on how fermentation, feminist theory, and cultural studies can change the way we think about our relationship to (microbial) foods.

NANCY HOALST-PULLEN is a professor of geography at Kennesaw State University and received her PhD in geography from the University of Colorado Boulder. She enjoys the art and science of homebrewing Belgian- and German-style ales, travels the world to meet people in the beer industry, and delivers experiential and applied talks on the geographies of beer. She is the coauthor of National Geographic's *Atlas of Beer*, a smart, comprehensive book with stunning photography, great storytelling, and intriguing destinations for those who love and appreciate beer and beer stories from around the world.

CHRISTOPHER R. HOLTKAMP, AICP, is an assistant professor in the Plant and Earth Science Department at the University of Wisconsin–River Falls. His research interests include place identity, social capital, and urban and rural development.

BRENDAN L. LAVY is an assistant professor in the School of Earth, Environmental, and Marine Sciences at the University of Texas Rio Grande Valley. He has a PhD and MS in geography from Texas State University and a BA in anthropology from the University of North Texas. He is a human-environment geographer with research interests in environmental resource management, disaster reconstruction and recovery, and urban sustainability and resilience.

SHADI MALEKI is a PhD student in the Department of Geography at Texas State University. She holds an MS in sustainability from Texas State University, an MA in international relations from the University of Perugia, Italy, and a BA in mass communications from the same institution. Her research interests include urban sustainability, geographies of children, and community development.

INNISFREE MCKINNON is an assistant professor of geography at the University of Wisconsin–Stout. Her research focuses on urban-rural interactions, political ecology, and critical geographic information systems. She uses a variety of methods, including qualitative methods, visual ethnography, and GIS.

ANDY MURRAY is a PhD candidate in sociology and a graduate student researcher with the Science & Justice Research Center at the University of California, Santa Cruz. His dissertation research focuses on community-based synthetic biology, fermentation, and bioengineering. His research interests span science and technology studies, health and medicine, bioengineering, multispecies studies, and agrifood studies.

COLLEEN C. MYLES is an associate professor in the Department of Geography at Texas State University in San Marcos. She has a PhD in geography and an MS in community development from the University of California, Davis. She is a rural geographer and political ecologist with specialties in land and environmental management; (ex)urbanization; (rural) sustainability and tourism; wine, beer, and cider geographies (aka "fermented landscapes"); and food/agriculture (urban, periurban, and sustainable).

JOHN OVERTON is a geographer at the School of Geography, Environment and Earth Sciences at Victoria University of Wellington, New Zealand. He has had a long-standing interest in the geography of wine and the way place has come to be an essential element in the global wine industry. He has conducted research on the wine industry in New Zealand, Australia, Chile, and South Africa.

MARK W. PATTERSON is a professor of geography at Kennesaw State University. He has cowritten and edited two books on beer, including National Geographic's *Atlas of Beer* and a forthcoming edited volume, *The Geography of Beer*, volume 2, due out in 2020.

ERIC SARMIENTO is an assistant professor at Texas State University. His interests include urban development, cultural geography, political ecology, economic geography, and social theory.

RUSSELL WEAVER is an associate professor in the Department of Geography at Texas State University. His current research interests include community economic development and collective action, particularly in the U.S. Appalachian region.

ELIZABETH YARBROUGH is a geographer and holds a master's in applied geography and a BS in resource and environmental studies from Texas State

University. Her interests include local, sustainable food production, home fermentation, and community networks.

PAUL ZUNKEL is an assistant professor of earth science in the Department of Physical Sciences at Emporia State University in Emporia, Kansas. He has a PhD in geography from Texas State University. His research interests include U.S. severe weather human-environmental interaction and the geography of wine, beer, and spirits.

Index

Page numbers in italics indicate illustrations. Page numbers appended with a "t" indicate tables.

bacteria: cacao beans and, 210; colony-forming units of, 267; in fermentation, xix, 17n4, 233, 256; in food preservation, 3; human interaction with, 268–69, 305, 340; from humans, 260–61, 262–63; modification of, 279; negative attitudes toward, 8, 230, 266; positive attitudes toward, 8, 15–16, 241, 266; in scobys, 221, 223–25

Balance Your Brew (listserv), 223

Banks Peninsula (New Zealand), 59, 63, 66, 67–68, 69–70, 71, 74–75

Barad, Karen, 268

Bardstown KY, *40*

barley, 86, 128, 131, 145

barrels, oak, 39, 86, 87, 101

bars (chocolate products): craft, 180; distributed farming model and, 191, 194; flavor norms for, 183–84, 214–15n5; mass-marketed, 185; quality issues in, 211, 212, 215nn10–11; single-estate model and, 191, 194, 209; snap and, 195; in tastings, 185, 188, *188t*, 201, *202–3t*, 204–5, 207–8, 215n9, 215n13; *terroir* in, 182; tourism and, 198

bars (drinking establishments), 31, 40, 138

Barton 1792 (bourbon), *40*, 94

Bateson, Gregory, 296n12

Battle of Puebla (1862), 144

Bavaria, 133, 140, 147n11

bean-to-bar chocolate: cost of making, 190; definition of, 214n2; farm-to-bar chocolate compared to, 211; growth of, 180, 182; potential for, 213–14; production models of, 194; quality of, 201, 204, 212; regional flavors in, 215n5; research methods for studying, 175–76, *188t*; in tastings, 188. *See also* bars (chocolate products); cacao and cacao beans; farm-to-bar chocolate

beech trees, 133

beer: Austrian immigrants and, 140; bacteria and, 266; Bloody Marys and, 160, *162*, 163, 167; British immigrants making, 128–30, 131–32; as chaser, 160, *162*; cider compared to, 110; craft production of, 285, 286, 288–89, 295n3; development in production of, 5–6; in England, 131; German immigrants and, 132–33, 134–36, 137, 140, 145; indigenous groups making, 127–28; influences on, xx; Irish immigrants and, 137–39, 145–46; laws on, 147n9; local identity and, 23, 31, 32; Mexican Americans and, 142–44, 146; in Mexico, 139–42; microbiology and, 281; popularity of, 127, 135; sales of, 24; tourism and, 39–41; yeast and, 17n2, 147n11, 283, 342

beerscapes, 127–28, 132, 144–46, 340

Benitez, Diego, 143

Bernardini, Georg, 184, 201, 215n9

beverages, carbonated, 133, 134, 227, 242–43

The Big Book of Kombucha (Crum and LaGory), 220

Big Island: as cacao bean source, 175, *177*, 216n14; cafés in, *198*; chocolate production in, *181*, 185, 196, *202–3t*, 204; history of chocolate industry on, 174; tourism in, *195*, 199

biodiversity, 5, 24–25, 119, 271n8

bioengineering, 276, 277–78, 280–81, 283–84, 286, 288–89, 290, 292–93, 294–95, 296n10. *See also* biology: synthetic; metabolic engineering

biofuels, 280, 282

biohacking, 275, 276, 287–89, 291, 296n13

biological specimens, 263–64

biology: cellular, 277; community, 287–91, 294; debates within field of, 295n4; European view in, 333n1; evolutionary, 320–21, 325; fermentation and, 273; molecular, 7, 277; synthetic, 274, 275, 276, 278, 280, 287–90, 342. *See also* bioengineering; metabolic engineering

biomedicine, 315

biopharmaceuticals, 279, 280

biopolitics, 17n5, 303–4, 313, 315, 341. *See also* microbiopolitics

bioreactors, 286–87

biosafety, 264, 271n9

biotechnology: as controversial topic, 276; fermentation as, 273–74, 277, 278–79, 288–89, 290, 340; history of, 5, 277–78, 292; regionalism and, 313

Bittenbender, H. C. "Skip," 174, 185

Black Canyon Distillery, 102

modity fetishism and, 199–200, 275–76; *craic* as, 137, 146, 147n8; exchange values of, 275; ferments as, 309; kombucha as, 307; landscape as, 26, 43, *44*; local culture in, 113; place as, 36; *terroir* and, 59; whiskey as, 89

commodity chains, 182, 211

community(ies): alternate food systems and, 232, 244–47, 248; drinking establishments in, 31, 145; group belonging and, 232–34; homebrewing and, 220, 225–26, 235, 238, 240; laboratories in, 287–91, 294; local food movements and, 16; Old World, 230; public good and, 41–42, 45; research in, 315; sense of place and, 28, 103–4, 341

community-supported agriculture (CSA) programs, 232

complexity, 264, 265–66, 267

computer science, 334n3

concentrated animal feeding operations (CAFOs), 270–71n7

conceptual deepening, 17–18n7

conceptual fluidity, 286

conching, 180, 207

condiments, 154, 166–67

conglomerates, 92, 143, 144

connotation and denotation, 159

consumers: agritourism and, 196, 198–99; community activities and, 104–5; connecting by consumption, 44; cultural differences among, 215n5; fermented foods and, 310, 331; genetic modification and, 284; health concerns of, 222, 230; historical focus of, 92; ignorance of, 199–200; of landscape, 31, 77; local focus of, 21, 22–24, 27–28, 32, 47n1, 102, 103, 164, 334n9; place important to, 94; production knowledge of, 42; production scales and, 118; quality variations and, 212

consumption: actor-network theory (ANT), 114; of alcohol, 31, 147n9; of craft products, 180; of fermented products, 16, 38–39; of GMOs, 284; influences on, 310; of kombucha, 242; of landscape, 13, 25–26, 34; localism and, 159, 166; microbes and, 303; production and, 153,

154, 292; of raw ferments, 312–13; sense of place and, 26–27, 44; of wine, 36, 63

contamination, 258–59, 260–61, 264, 266, 288–89, 303, 306

content analysis of websites, 95–97, *96t*, 98

control: of fermentation, 7, 287, 309, 340; of fermenting containers, 206–8; of food systems, 312–13; of human bodies, 17n5; in kitchens, 258–60; in laboratories, 258–60; of microbes, 7, 17n2, 17n5, 262, 266–67; power and, 266–67; purity and, 266–67; theory of, 258–60

Cooper, Bob, 174, *195*

Cooper, Pam, 174

coopetition (cooperative economics), 26, 121

Corbans Wines, 65, 74

corn: in adjunct lagers, 147n10; as agricultural product, 86, 89; in beer, 128, 131; in bourbon production, 39, 86–88, 89, 90, 102; land dedicated to, 97, *99*, 100–101, *100t*; as local product, 85, 98, 102; in whiskey production, 89

Côte d'Ivoire, 215n6

county subdivisions in bourbon production, 97, 100–101, *100t*

couverture, 196, 214n2

covariance between fermentation and landscape change, 319–20, 327–28, 329, 330

Craft Bourbon Trail, 38

craft brewers: impact of, 47n3; marketing by, 39–41

craft brewing: bioengineering compared to, 281–86; growth of, 24; history of, 142–44; impact of, 26, 42; local emphasis by, 23; milk substitute production as, 283–86; small-scale, 23; in Wisconsin, 32

craft producers and production: craft chocolate in, 180; criteria for, 285–86, 296n11; definition of, 47n2; fermentation enterprises as, 330–31, 332; growth of, 21, 23–24; homebrewing and, 47n1; impact of, 22, 26–27, 28–30, 42; large-scale, 109–10; local focus of, 23, 24–25; rural landscapes and, 24–25, 43; sense of place and, 27–28, 42–43, 45–46; urban landscapes and, 26, 43–44; in Wisconsin, 30–34

fermentation (word), 276

fermentation fetishism, 275–76, 283

fermentation technology (zymotechnology), 277, 278, 287, 295n6

fermented landscapes: chocolate production and, 176, 178–80, 213–14; cider production and, 110, 112–13; coopetition and, 120–21; craft producers creating, 22, 45; definition of, 3, 10, 11; enterprise and, 327–32; as explanatory, 330–31; as framework, xix, xxi, 11, 15, 16, 337–39, 341; as fruitful, 327–29; as illuminating, 325–26; kombucha production and, 247–48; migration and, 128, 145–46; overview of, xx, 343; political issues and, 15–16, 301–2; power and, 16–17, 315; as probative, 330–31; research and, 304, 311, 313–14, 314–15, 341–43; as research paradigm, 319–20, 331, 333; in Sierra Nevada foothills, 13–14; suburban sprawl and, 327; urban issues and, 332; in wine regions, 58, 75–76, 77. *See also* landscape change

fermentos, 8–9, 15–16, 230, 232, 234, 243. *See also* brewers: beer; brewers: kombucha; brewers: wine

ferments: pasteurized, 302, 314, 341; raw, 302, 306, 308, 310–13, 314–15

fertilizers, 191, 231

Fine Chocolate and Cacao Institute, 214n4

flavor, 176, 178–79, 182, 183–84, *186t, 202–3t,* 214–15n5, 215n9, 215n11. *See also* off-flavors

floral flavor, 214–15n5

FLs. *See* fermented landscapes

focus groups, 237–38

föhn, 63

food: acceptability of, 262, 270–71n7; alcohol and, 6–7; community and, 232–34; consumer demand and, 331, 334n9; functional, 222, 241, 307; as garnish, 162, 164; handmade, 262–63; health and, 308; illness and, 270n4; labor and, 262, 270–71n7; local, 31, 34, 159; microbes and, 5–6, 12, 260–63, 266, 269, 306, 310–11; in modern diet, 17n4, 308; preservation of, 261; production of, 280, 283–84, 285, 303–4, 312–13; regulation of, 306–7, 315;

research on, 255–58, 270n2; safety of, 4, 7, 260, 263–64, 303–4, 306, 315; significance of, 153–54, 159; substitutes for, 273, 280, 283, 290; tastings and, 265–66. *See also* bars (chocolate products); bean-to-bar chocolate; cheese; chocolate; farm-to-bar chocolate

food, fermented: as acquired taste, 17n6; benefits of, 5–6, 233–34, 301–2; chocolate as, 176; connections from, 269; ecological acts and, 15–16; as enlivened, 7–8, 241, 307; handmade, 262; health and, 236–37, 242, 308; home-brewing of, 309–10; humans coevolving with, 8–9; political acts and, 15–16; potential for, 310–12; for preservation, 3–4, 230; safety and, 306–7; sense of place and, 15–16; significance of, 229–31, 340

foodies, 34, 210, 312

Foodland, *197*

food sovereignty, 312–14

food subjectivities, 310–12, 315

food systems: alternative, 22, 29–30, 229–30, 231–32, 240, 243–44, 248, 292; class issues in, 29–30; corporate control of, 312; industrial, 309; inequalities in, 311–12; local, 15, 16, 24; Pasteurianism and, 303, 306–7, 308; post-Pasteurianism and, 307; potential for, 310, 313–14, 315; race issues in, 29–30

foodways, 4, 313–14

Forbes, Stephen, 103

Foucault, Michel, 310–11, 315

fragrance, 279, 280, 284

France and French influence, 87, 130, 140, 153, 159, 271n8

Frank, Günther, 222

French Farm Winery, 63, 67, 74

fruitful paradigms, 320, 325, 328–29, 333, 334n5

fruits, 4, 10, 112, 115, 127, 199

fruity flavor, 183, 214n5

fuels, 273, 280, 282

functional proximity, 71

fungi, 3, 8, 10, 15, 233, 270

Gadsden Purchase (1853), 141

73–74, 78n3; sense of place and, 35–36; in
Texas, 36–38, 43, *44*; in Wisconsin, 32
Wisconsin, 30–34, *33*
Wittgenstein, Ludwig, 257
women, 29–30, 94, 138, 146n7
Wonder Drink, 222
Wonder-Pilz, 219
woody flavor, 183
Worcester MA, 137–38
Worcestershire, England, 110, *111*, 112
workers, unauthorized, 147n14
World War II, 221
wounds, flesh, 262

yeasts: in alcoholic beverages, 6; in fermentation,
xix, 9, 133, 135, 147nn10–11, 206, 210, 233, 256,
280; human interaction with, 340, 342; from
humans, 262; knowledge about, 4, 17n2;
modification of, 279, 282–83, 284–85, 287,
288–89; positive attitudes toward, 241; in
research, 259; in scobys, 221, 223–25; transport
of, unintentional, 133
Yellowstone (bourbon brand), 92
Yuengling, 134

Zbikiewicz, Henry, 157
Zizania texana, 228–29
zymotechnology (fermentation technology), 277,
278, 295n6
zymotic theory of disease, 295n5
zymurgeography, xxi. *See also* biotechnology;
fermented landscapes

Milton Keynes UK
Ingram Content Group UK Ltd.
UKHW012152230424
441302UK00020B/9

9 781496 207760